Lecture Notes
in Economics and
Mathematical Systems

Managing Editors: M. Beckmann and W. Krelle

344

Klaus Neumann

Stochastic Project Networks

Temporal Analysis, Scheduling and Cost Minimization

Springer-Verlag

Berlin Heidelberg New York London Paris Tokyo Hong Kong

Managing Editors

Prof. Dr. M. Beckmann
Brown University
Providence, RI 02912, USA

Prof. Dr. W. Krelle
Institut für Gesellschafts- und Wirtschaftswissenschaften
der Universität Bonn
Adenauerallee 24–42, D-5300 Bonn, FRG

Author

Prof. Dr. Klaus Neumann
Institut für Wirtschaftstheorie und Operations Research
Universität Karlsruhe
Kaiserstr. 12, D-7500 Karlsruhe 1

ISBN-13:978-3-540-52664-3   e-ISBN-13:978-3-642-61515-3
DOI: 10.1007/978-3-642-61515-3

2142/3140-543210 – Printed on acid-free paper

Preface

Project planning, scheduling, and control are regularly used in business and the service sector of an economy to accomplish outcomes with limited resources under critical time constraints. To aid in solving these problems, network–based planning methods have been developed that now exist in a wide variety of forms, cf. Elmaghraby (1977) and Moder et al. (1983).

The so–called "classical" project networks, which are used in the network techniques CPM and PERT and which represent acyclic weighted directed graphs, are able to describe only projects whose evolution in time is uniquely specified in advance. Here every event of the project is realized exactly once during a single project execution and it is not possible to return to activities previously carried out (that is, no feedback is permitted).

Many practical projects, however, do not meet those conditions. Consider, for example, a production process where some parts produced by a machine may be poorly manufactured. If an inspection shows that a part does not conform to certain specifications, it must be repaired or replaced by a new item. This means that we have to return to a preceding stage of the production process. In other words, there is feedback. Note that the result of the inspection is that a certain percentage of the parts tested do not conform. That is, there is a positive probability (strictly less than 1) that any part is defective. Hence the return to the preceding stage of the production process generally occurs with a probability less than 1. Furthermore, the activities "repair the part" and "order a new item" may be carried out several times. Aside from the planning and supervision of production processes, many R&D projects possess certain new properties, some of which have been mentioned above, that cannot be treated within the framework of "classical" project networks.

To deal with these more general projects, **stochastic project networks** have been introduced. They possess more general arc weights, several different node types (containing stochastic elements), and cycles. Again the activity–on–arc representation known from CPM and PERT is used. Stochastic project networks are also called **GERT networks**, a name introduced by Pritsker who has done much pioneering work in this area.

This monograph presents the state of the art of temporal analysis and cost minimization of projects, as well as project planning under limited resources where the projects are modelled by GERT networks. For a more detailed exposition of certain advanced topics, the reader is referred to the extensive bibliography. We focus upon analytical methods only. Thus, we do not deal with the method of simulation. The monograph by Neumann and Steinhardt (1979a) gives a comprehensive survey of the results known by 1979 on the time–oriented evaluation of GERT networks. The chapters on temporal analysis in the present book represent an improved and expanded version of the essential parts of Neumann and Steinhardt (1979a), with several new results included. The chapters on cost minimization and scheduling with GERT precedence constraints are completely new. This monograph does not contain any applications. Numerous examples of applications of different types of GERT networks can be found in Neumann and Steinhardt (1979a, 1979b) and in Foulds and Neumann (1989). The book will be of interest to graduate students and researchers in project planning and scheduling, machine scheduling, and stochastic networks. More generally, the book is useful to potential readers who want to study stochastic models in operations research.

The contents of this book are as follows: Chapter 1 defines GERT networks and deals with some structural questions. The basic concepts and results from the temporal analysis of GERT networks make up chapter 2. Chapters 3 and 4 are devoted to the evaluation of special classes of GERT networks with regard to time planning. Chapter 3 discusses GERT networks all of whose nodes have exclusive–or entrance (so–called EOR networks), which can be associated with Markov renewal processes. Results from the theory of those stochastic processes can be exploited for the evaluation of EOR networks. Chapter 4 focusses on GERT networks that may contain all kinds of GERT nodes and can be reduced to EOR networks. Chapter 5 studies single–machine scheduling problems with precedence constraints given by a GERT network. This corresponds to the case where a single resource of capacity 1 is required to carry out each activity of the respective project. Single–machine min–sum scheduling problems with general GERT precedence constraints are solved here by dynamic programming. Polynomial algorithms are presented for scheduling problems with EOR precedence constraints, where the weighted expected flow–time or maximum expected lateness are to be minimized. Chapter 6 discusses the cost minimization of projects modelled by EOR networks, which leads to Markov renewal decision processes and can thus be reduced to a stochastic dynamic programming problem. Chapter 7 deals with cost and time minimization of so–called decision project networks, which represent a class of project networks showing some affinity to GERT networks.

The monograph assumes a knowledge of elementary mathematical programming, probability theory, and Markov chains. Some basic concepts from the theory of graphs and networks, Markov chains and Markov renewal processes, deterministic scheduling, and dynamic programming are summarized before they are needed. The coverage of these topics makes the monograph self–contained. Proofs of certain of the theorems are omitted when they are not of sufficient intrinsic interest or can be found in literature.

I would like to thank Matthias Bücker and Philipp Derr who read the entire manuscript and suggested several improvements. I am especially indebted to Mrs. Christa Otto for the careful typing and patient retyping of the text in T^3. The drawings were done by Bernd Hornung by means of **Autosketch.**

Conweiler, The Black Forest   Klaus Neumann
December 1989

Contents

List of Symbols

Miscellaneous

:=	Equal by definition
■	End of proof or of algorithm
$\lfloor x \rfloor$	greatest integer $\leq x$

Sets

$\mathbb{R}$	Set of real numbers
$\mathbb{R}_+$	Set of nonnegative real numbers
$\mathbb{R}_{++}$	Set of positive real numbers
$\mathbb{R}^n$	Set of n-tupels of real numbers
$\mathbb{N}$	Set of positive integers (natural numbers)
$\mathbb{N}_0 := \mathbb{N} \cup \{0\}$	Set of nonnegative integers
$\emptyset$	Empty set
$|M|$	Number of elements of a finite set M
$N \subseteq M$	N is a subset of M
$N \subset M$	N is a proper subset of M
$M \backslash N := \{a \in M \mid a \notin N\}$	Difference of the sets M and N
$\mathbb{P}(M)$	Power set (set of all subsets) of M
$\mathring{\mathbb{P}}(M) := \mathbb{P}(M) \backslash \{\emptyset\}$	Set of all nonempty subsets of M
$f : M \rightarrow N$	Mapping (function) of M into N
$O(f(n))$	Landau's symbol "big oh" : For $f, g : \mathbb{N} \rightarrow \mathbb{R}_+$ it holds that $g(n) = O(f(n))$ if there are a constant $c > 0$ and a positive integer n_0 such that $g(n) \leq cf(n)$ whenever $n \geq n_0$

Probability

$P(A)$	Probability of random event A
$P(A \mid B)$	Conditional probability of event A given B
$P(A \cup B)$	Probability of event "A or B"
$P(A \cap B)$	Probability of event "A and B"
$\mathbb{E}(X)$	Expected value of random variable X
$\mathbb{E}(X \mid A)$	Conditional expectation of random variable X given event A

Directed Graphs and Networks

V	Node set of a directed graph or network (set of project events)
E	Arc set of a directed graph or network (set of activities)
$<i,j>$	Arc (activity) with initial node i and final node j
R	Set of sources
S	Set of sinks
$\mathcal{P}(i)$	Set of predecessors of node i
$\mathcal{S}(i)$	Set of successors of node i
$\mathcal{R}(i)$	Set of nodes reachable from node i
$\mathcal{\dot{R}}(i) :=$	$\mathcal{R}(i)\backslash\{i\}$
$\mathcal{\bar{R}}(i)$	Set of nodes from which node i is reachable

GERT Networks

Ω	Set of all possible project realizations (sample space)
p_{ij}	Execution probability of activity $<i,j>$
D_{ij}	Duration of activity $<i,j>$
F_{ij}	Distribution function of D_{ij}
$\mathcal{A}_i$	Random event "node i is activated"
$\bar{\mathcal{A}}_i$	Random event "node i is not activated"
T_i	Time of activation of node i
$q_{R'}(q_{S'})$	Probability that at least the sources i∈R' (sinks i∈S') are activated
$\hat{q}_{R'}(\hat{q}_{S'})$	Probability that exactly the sources i∈R' (sinks i∈S') are activated
q_i	Probability that (at least) node i is activated
Y_U	Activation function of set U
$y^U_{U'}$	Activation distribution of set U' with respect to set U
$Y(S):=\{Y_{S'}\|S'\in\mathbb{P}(S)\}$	Family of activation functions of sink set S
Y_j	Activation function of node j
Y_{ij}	Activation function of node j given source i
z_j	Activation number of node j
z_{ij}	Activation number of node j given source i
D	Project duration
D^{skip}	Skipping project duration

GERT Scheduling (Chapter 5)

C_{ij}	Completion time of activity $<i,j>$
$f_{ij}(t)$	Cost incurred when activity $<i,j>$ is completed at time t
w_{ij}	Weighting factor of activity $<i,j>$
o	Operation
L_o	Lateness of operation o
$\mathcal{O}$	Set of operations
$\mathcal{O}_f$	Set of final operations
$\mathcal{q}$	List schedule or precedence schedule
$\mathcal{Q}$	Set of precedence schedules
$\mathcal{Q}_o$	Set of all precedence schedules whose last element is the final operation o
$\mathcal{X}$	Set of all feasible arc–node sets
ϕ	Scheduling policy
H^*	Minimum total cost function

Decision Networks (Chapter 7)

χ_i^-	Entrance characteristic of node i
χ_i^+	Exit characteristic of node i
u_i	Node variable
w_{ij}	Arc variable
w	Project realization (deterministic action)
$\mathcal{E}$	Set of admissible project realizations
c_{ij}	Cost of activity $<i,j>$
π	Randomized action
ψ	Policy (sequence of randomized actions)
d_{ij}	Duration of activity $<i,j>$
t_i^w	Activation time of node i in project realization w
t_i^e	Earliest possible activation time of node i

Chapter 1 Basic Concepts

In chapter 1 we first summarize the most important concepts from the theory of graphs and project networks which are needed in what follows. For additional background we refer to Elmaghraby (1977), Lawler (1976), and Neumann (1987a, 1987b). After that, the concept of a GERT network is introduced and some basic assumptions and results are stated.

1.1 Directed Graphs and Project Networks

A **directed graph** G is given by a set E of **directed edges** or **arcs**, a nonempty set V of **vertices** or **nodes**, and two so–called **incidence mappings** which assign an **initial node** (or tail) and a **final node** (or head) to each arc. The initial node and final node of an arc are also called its **endnodes**. Arcs are usually illustrated by arrows and nodes by points (see Fig. 1.1.1).

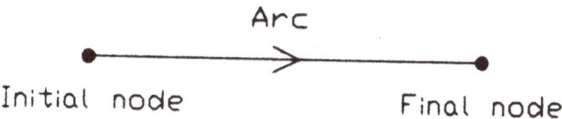

Figure 1.1.1

In what follows, we can restrict ourselves to the case where the directed graph does not contain parallel arcs, that is, arcs that share the same initial and final node (see Fig. 1.1.2). Then each arc is uniquely specified by its initial node, say i, and its final node, say j, and we use the symbol <i,j> for that arc (compare Fig. 1.1.3). Also, we assume that the directed graph does not contain loops, that is, arcs whose initial node coincides with the final node.

Figure 1.1.2   Figure 1.1.3

For a directed graph with node set V and arc set E we use the symbol <V,E>. Since each arc <i,j> can be identified with a pair (i,j)∈V×V, the arc set E will be considered a subset of V×V in what follows. In this monograph, we will only deal with finite graphs, that is, the sets V and E are finite.

Given an arc <i,j>, j is called a **successor** of i, and i is called a **predecessor** of j. By $S(i)$ we denote the set of successors of a node i and by $P(i)$ the set of predecessors of node i. A **source** is a node which has no predecessors and a **sink** is a node without successors (see Fig. 1.1.4). The set of the sources of a directed graph is designated by R and the set of sinks by S. A node that possesses neither predecessors nor successors is called **isolated.**

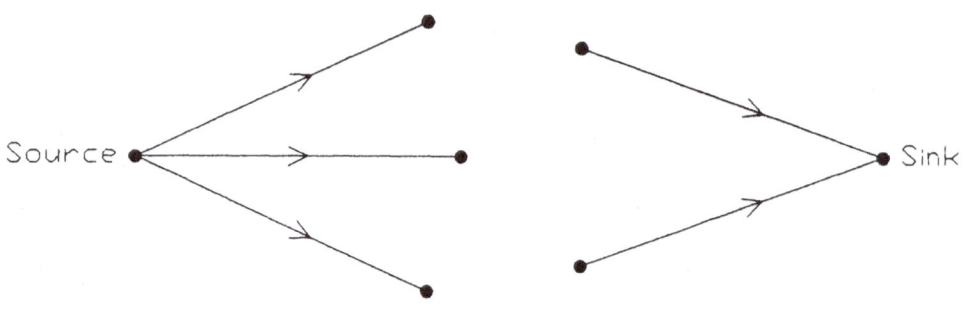

Figure 1.1.4

A directed graph G'=<V',E'> is called a **subgraph** of G=<V,E> if V'⊆V and E'⊆E. If V'⊆V, then the subgraph of G **induced by** V' has the node set V' and contains all arcs <i,j> from G such that both i and j are in V'. We also speak of an **induced subgraph** G' of G=<V,E> if G' is induced by some set V'⊆V. G'=<V',E'> is called a **proper subgraph** of G=<V,E> if V'⊂V or E'⊂E. Fig. 1.1.5 shows a directed graph G, a proper subgraph of G, and the subgraph of G induced by the node set {1,2,3}.

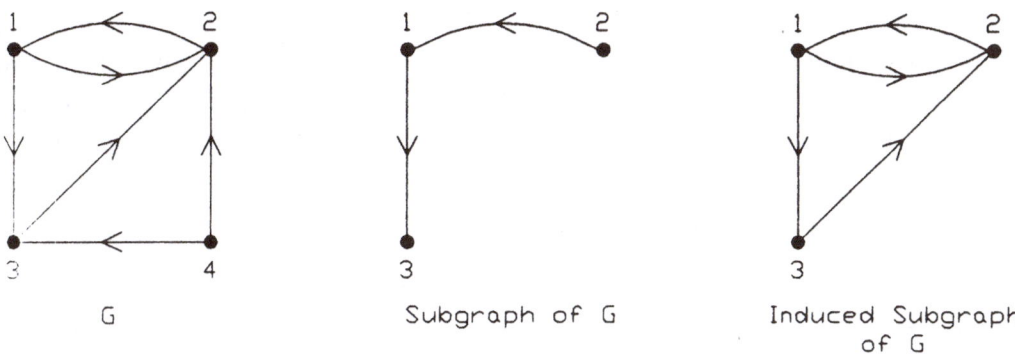

G   Subgraph of G   Induced Subgraph
                                                     of G

Figure 1.1.5

A sequence of arcs, where the final node of any arc coincides with the initial node of the next arc is called a **walk.** In Fig. 1.1.6 there is a walk **from** i_0 **to** i_4 consisting of the arcs $\langle i_0,i_1\rangle$, $\langle i_1,i_2\rangle$, $\langle i_2,i_3\rangle$, $\langle i_3,i_1\rangle$, $\langle i_1,i_2\rangle$, and $\langle i_2,i_4\rangle$. We use the symbol $\langle i_0,i_1,i_2,i_3,i_1,i_2,i_4\rangle$ for that walk. i_0 is called the **initial node** and i_4 is the **final node** of that walk. If the final node is different from the initial node, the walk is called **open,** otherwise it is **closed.** An open walk $\langle i_0,i_1,\ldots,i_r\rangle$ in which all nodes $i_0,i_1,\ldots,i_r$ are distinct is called a **path,** and a closed walk $\langle i_0,i_1,\ldots,i_r\rangle$ in which all "intermediate" nodes are distinct (that is, $i_r=i_0$ and $i_k\neq i_1$ for $k\neq1$; $k,l=1,2,\ldots,r$) is called a **cycle.** A directed graph which does not contain any cycles is said to be **acyclic.** In Fig. 1.1.6, $\langle i_0,i_1,i_2,i_4\rangle$ represents a path and $\langle i_1,i_2,i_3,i_1\rangle$ is a cycle.

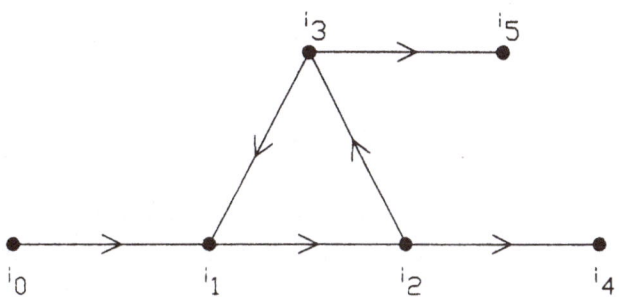

Figure 1.1.6

If, in a sequence of arcs, not all arcs have the same direction (that is, one of the two endnodes of any arc coincides with one of the endnodes of the next arc), we speak of a **semiwalk**. In Fig. 1.1.6, the sequence of arcs $\langle i_0, i_1 \rangle$, $\langle i_3, i_1 \rangle$, $\langle i_3, i_5 \rangle$ represents a semiwalk with the **endnodes** i_0 and i_5.

A node j is called **reachable** from a node i if there exists a walk (and thus a path) with initial node i and final node j. A node is also considered to be reachable from itself. The set of the nodes that are reachable from node i is designated by $\mathcal{R}(i)$, and the set of those nodes from which node i is reachable is designated by $\overline{\mathcal{R}}(i)$. Moreover, we define

$$\mathring{\mathcal{R}}(i) := \mathcal{R}(i) \backslash \{i\}$$

$$\mathcal{R}(U) := \bigcup_{i \in U} \mathcal{R}(i), \ \overline{\mathcal{R}}(U) := \bigcup_{i \in U} \overline{\mathcal{R}}(i) \quad \text{for } \emptyset \neq U \subseteq V$$

For the directed graph in Fig. 1.1.7, we have $\mathcal{R}(1) = \{1,2,3,4\}$, $\mathring{\mathcal{R}}(1) = \{2,3,4\}$, and $\overline{\mathcal{R}}(1) = \{1,2,3,5\}$.

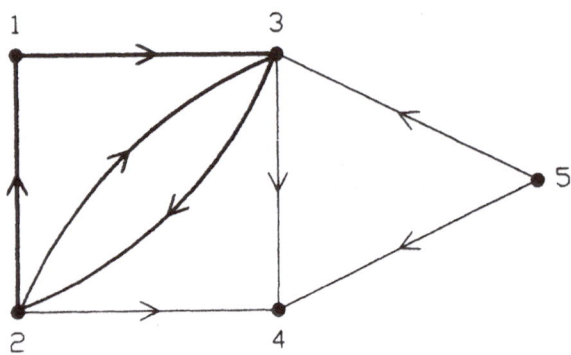

Figure 1.1.7

Two nodes i and j are called **joined** if there is a semiwalk with the endnodes i and j. Any node is considered to be joined to itself. In Fig. 1.1.7 every two nodes are joined.

A directed graph G is said to be **weakly connected** if any two nodes of G are joined. G is called **strongly connected** if for any two nodes i and j of G, i is reachable from j and j is reachable from i. A **weak component** (or respectively **strong component**) of G is a maximal weakly (or respectively strongly) connected subgraph of G, i.e., it is not a proper subgraph of any other weakly (or respectively strongly) connected subgraph of G. The digraph in Fig. 1.1.7 is weakly connected. The subgraph induced by the node set $\{1,2,3\}$ (indicated by darker arrows) is a strong component. Moreover, the two subgraphs consisting of the single nodes 4 and 5, respectively, represent strong components. A strong component different from an isolated node is called a **cycle structure**. A cycle structure represents, figuratively speaking, a maximal set of connected cycles.

(1.1.1) We introduce an order $\preceq$ in the node set V of a directed graph by

$$i \preceq j \text{ exactly if } i=j \text{ or } j \in \mathcal{R}(i) \backslash \mathcal{R}(i) \text{ for } i \neq j$$

$i \prec j$ (in words, "i **before** j" or "j **behind** i") means that $i \preceq j$ and $i \neq j$ ($\prec$ is a strict order in V). Now let $V = \{1, 2, \ldots, n\}$. We say that the nodes of a directed graph are **topologically ordered** if

$$i \prec j \text{ implies } i < j \quad (1 \leq i, j \leq n)$$

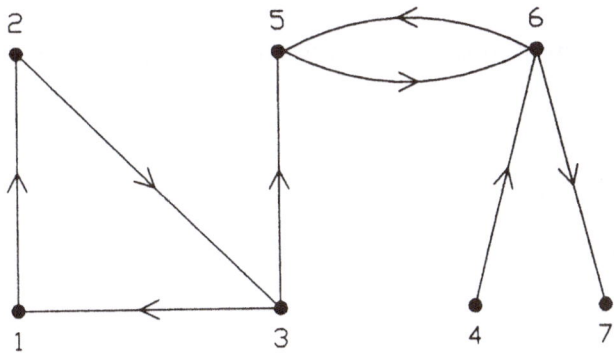

Figure 1.1.8

In the latter case, we have $j > i$ for all $j \in \mathcal{S}(i)$ and $k < i$ for all $k \in \mathcal{P}(i)$ provided that i and j or i and k, respectively, do not belong to one and the same cycle structure. For example, the nodes of the directed graph in Fig. 1.1.8 are topologically ordered. The nodes of an acyclic directed graph can be ordered topologically and the cycle structures of a directed graph can be found each in $O(|E|)$ time (cf. Neumann (1987a), section 6.2.3). Thus the nodes of a directed graph with cycles can be ordered topologically in $O(|E|)$ time as well.

(1.1.2) We can also introduce a strict order in the set of the cycle structures of a directed graph G. Let C_1 and C_2 be two disjoint cycle structures of G. Then $C_1 \prec C_2$ (in words, "C_1 **before** C_2" or "C_2 **behind** C_1") precisely if the nodes of C_2 are reachable from the nodes of C_1. Similarly, a node k is **before** a cycle structure C (or C is behind k) and another node j is **behind** C (or C is before j) if $k \prec i$ and $i \prec j$ for each node i of C. In Fig. 1.1.8, the cycle $\langle 1,2,3,1 \rangle$ and the nodes 1 to 4 are before the cycle $\langle 5,6,5 \rangle$, whereas node 7 is behind $\langle 5,6,5 \rangle$. If, in addition to a cycle structure C of G, there are r−1 cycle structures $C_1, \ldots, C_{r-1}$ of G for which

$$C_1 \prec C_2 \prec \ldots \prec C_{r-1} \prec C$$

but there are no r cycle structures $C'_1, \ldots, C'_r$ with

$$C'_1 \prec C'_2 \prec \ldots \prec C'_r \prec C$$

then we assign the **rank** r to the cycle structure C. In Fig. 1.1.8, cycle $\langle 1,2,3,1 \rangle$ has rank 1 and cycle $\langle 5,6,5 \rangle$ has rank 2.

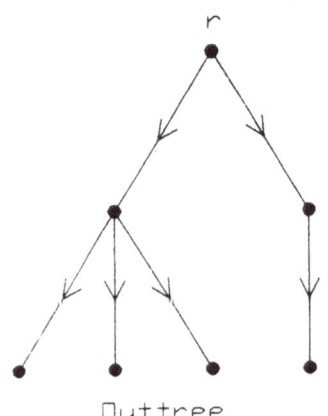

Outtree

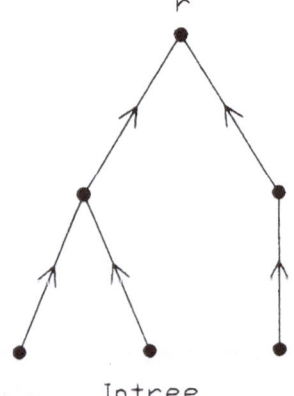

Intree

Figure 1.1.9

(1.1.3) A directed graph G is called a **directed tree rooted from node** r or an **outtree with root** r if r is the source of G and each node of G aside from the root has exactly one predecessor. G is called a **directed tree rooted to node** r or an **intree** if r is the sink of G and each remaining node has exactly one successor (see Fig. 1.1.9). A **directed forest** is a directed graph whose weak components are directed trees. In particular, a directed graph G is said to be an **outforest** if each node of G has at most one predecessor. Analogously, an inforest is defined. We note that each directed forest is acyclic.

A directed graph G=<V,E> is called **bipartite** if it contains only arcs <i,j> with i∈R and j∈S (see Fig. 1.1.10). Bipartite directed graphs are acyclic. If E={<i,j>|i∈R, j∈S}, the bipartite directed graph is said to be **complete**.

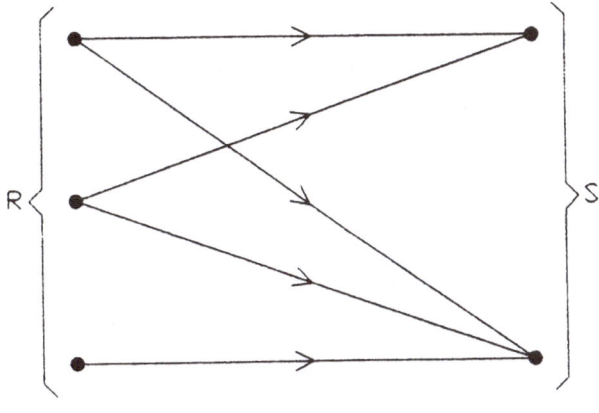

Figure 1.1.10

A directed graph G=<V,E> together with a mapping w:E→M, which assigns a weight $w_{ij} \in M$ to each arc <i,j> of G and where M is any set, is called a **weighted directed graph**. If the weights are real numbers, we define the **length** of a walk in G to be the sum of the weights of its arcs.

In the literature of graph theory, where terminology is quite unstandardized, there are different definitions of the concept "network". Mostly, each weighted directed graph is referred to as a network. We assume in addition that a network does not have isolated nodes because most weighted directed graphs in applications (for example, in project planning and control, shortest route problems, transportation problems, flows in networks) possess that property.

(1.1.4)   Definition.

A weighted directed graph without isolated nodes is called a **network**. We speak of a **network** N **with sources and sinks** if

(a)   N contains at least one source and at least one sink

(b)   Each node of N is reachable from at least one source and from each node at least one sink is reachable.

The reason for introducing the concept of a network with sources and sinks is that project networks represent networks with sources and sinks. Fig. 1.1.11 shows a network with sources and sinks where we have omitted the arc weights.

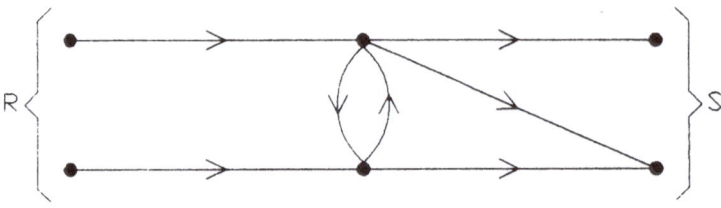

Figure 1.1.11

Each network contains at least one arc because it has at least one node and it does not possess isolated nodes. Each acyclic network represents a network with sources and sinks (proof by induction). On the other hand, a strongly connected network which represents a cycle structure has neither sources nor sinks.

(1.1.5) A **subnetwork** of a network N is a subgraph of N which does not contain isolated nodes and whose arcs have the same weights as the corresponding arcs in N. Similarly, the concept of an **induced subnetwork** is defined.

(1.1.6) A node i of a subnetwork N' of a network N is called an **entrance node** (or an **exit node**, respectively) of N' if i is a source (or a sink, respectively) of N or the final node (or initial node, respectively) of an arc in N not belonging to N'. In Fig. 1.1.12, nodes i_1, i_2 and i_3 are entrance nodes and j_1 and j_2 are exit nodes of N'. Node k is at the same time an entrance node and an exit node of N'. Owing to condition (1.1.4b), every subnetwork of a network with sources and sinks has at least one entrance node and at least one exit node.

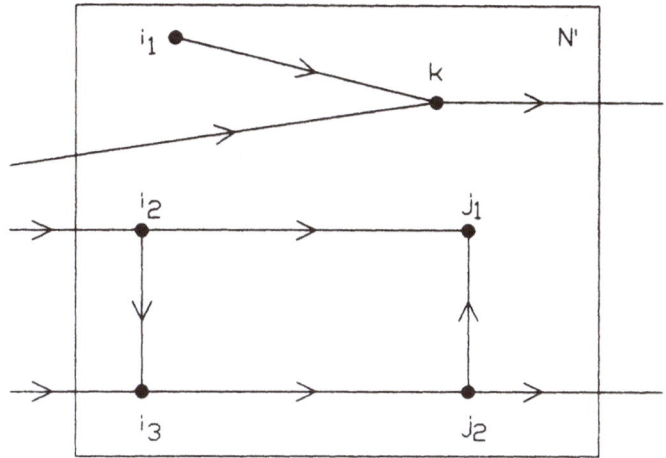

Figure 1.1.12

(1.1.7) An arc in a network N whose final node (or initial node, respectively) is an entrance node (or exit node, respectively) of a subnetwork N' of N but which does not belong itself to N' is called an **entrance arc** (or **exit arc**, respectively) of N'.

(1.1.8) A node i of a subnetwork N' of a network N is called a **source** (or a **sink**, respectively) of N' if no predecessor (or successor, respectively) of i belongs to N'. A source of N' represents a special entrance node, and a sink is a special exit node. In Fig. 1.1.12, i_1 and i_2 are the sources and k and j_1 are the sinks of N'. Note that i_3 and k are entrance nodes but no sources, and j_2 is an exit node but no sink.

(1.1.9) **Remarks.**
(a) Distinct sources (and distinct sinks) of a subnetwork N' are not reachable from one another "within N'".
(b) Let R' and S' be the set of the sources and the set of the sinks, respectively, of a subnetwork N'. Then $R' \cap S' = \emptyset$.
(c) Let N' be a subnetwork of a network with sources and sinks. If each entrance node of N' is a source of N' and each exit node of N' is a sink of N', then N' represents a network with sources and sinks if treated as a separate entity.

For the planning, scheduling and controlling of projects, **project networks**, also called **activity networks**, have been introduced. To this end, the project is first decomposed into certain nonoverlapping tasks called **activities**, each of which requires time and possibly resources for its completion (among the activities, we also count time delays such as delivery and waiting periods). The beginning of an activity and the termination of an activity are called **(project) events**. The next step is to establish the precedence relationships regarding the order in which the activities must be carried out and in which the events occur. Finally, one needs the **durations** of the individual activities (for the so–called time planning).

The construction of a corresponding project network (using the so–called **activity–on–arc representation**) is then as follows: An arc is assigned to each activity of the project in question whereby the duration of the activity corresponds to the weight of the respective arc. The initial node and the final node of an arc correspond to the two events "beginning of the associated activity" and "termination of the associated activity", also called "initial event" and "final event" of the respective activity. Contiguous arcs are assigned to activities that succeed one another immediately, where it may be necessary to introduce **dummy activities** (of zero duration) in order to represent the precedence relationships among the activities correctly (for details we refer to Elmaghraby (1977) and Neumann (1987b)).

An activity network which is assigned to a project in the above manner represents a network with sources and sinks. In particular, it has, possibly by introducing dummy activities, exactly one source (corresponding to the "beginning event" of the project), and it is acyclic. The well–known network techniques CPM and PERT use such project networks.

In what follows, we will synonymously employ the related concepts "activity" and "arc" on the one hand, "event" and "node" on the other hand, and thirdly "project" and "network". We also speak of the **activation of a node** if the corresponding project event occurs and of the **execution of an arc** or of a walk.

1.2 GERT Networks

The "classical" project networks used in the network techniques CPM and PERT are applicable only to those projects for which every activity and every event are realized exactly once during a realization of the respective project. In real–world projects (for example, R&D projects or in the planning and supervision of production processes), however, it can happen that certain activities are carried out only with a probability less than 1. Moreover, it is possible that an event occurs already at the time when not all but, say, only one of the activities leading into this event has been completed. Finally, it can happen that the project, during the course of its execution, will return to events that have already occurred before (in other words, feedback is allowed). As a consequence, there may be activities that are not carried out and activities that are carried out several times during a single realization of a project.

We now give an example which comprises the three particular properties just mentioned and which is borrowed from Neumann and Steinhardt (1979a), section 1.2. It deals with the development of a new product by an enterprise. Two research groups, designated A and B, work independently towards the development of one new product each, and each group carries out tests for the suitability of its product at the end of the development period. As a rule, if such a test proves the product to be not suitable, a product improvement followed by a new test ("feedback") is undertaken (the activity "product improvement" has, for example, a probablity of execution less than 1). If, however, the product is totally unsuitable, it is recommended to discontinue the development of this product. The entire development project is to be considered successful when one of the two research teams develops a suitable new product. The project will, on the other hand, be unsuccessful if both teams discontinue the development of their product.

The three properties mentioned above, which are not amenable to the "classical" project networks, can be treated through an extension of the weighting of the arcs and through the introduction of various kinds of nodes and of cycles in a project network with sources and sinks based on activity–on–arc–representation. The more general networks thus obtained are called **stochastic project networks** (since the evolution of the corresponding projects is no longer uniquely determined), or they are called **GERT networks,** a name introduced by Pritsker, who has done much research in this field.

Since GERT networks describe projects with stochastic evolution structure and stochastic activity durations, we have to establish a basic probability space. The sample space Ω (that is, the set of all possible outcomes of the random experiment in question, which consists of carrying out the underlying project) is identified with the set of all possible realizations of the project or, respectively, the associated network.

(1.2.1) Note that we have to distinguish between the concepts **project execution** and **project realization** (or, respectively, between **network execution** and **network realization**). A project execution corresponds to a performance of the basic random experiment, whereas a project realization represents an outcome of that experiment. Also, we have to distinguish between the concepts "(project) event" and "random event". The occurrence of a project event, however, represents a special random event in the case of a GERT network. We use the following

(1.2.2)   Notation.

The random event "node i is activated" is denoted by A_i and the random event "node i is not activated" by $\overline{A}_i$.

In what follows, we discuss the weighting of the arcs, the so-called initial distribution, and the node types of a GERT network. First we take a look at the **arc weights.** A weight vector $\begin{bmatrix} p_{ij} \\ F_{ij} \end{bmatrix}$ is assigned to each activity $\langle i,j \rangle$. The first component p_{ij} is the conditional probability that activity $\langle i,j \rangle$ is carried out given that its beginning event i has occurred (briefly called the **execution probability** of activity $\langle i,j \rangle$):

(1.2.3) $p_{ij} := P(\langle i,j \rangle$ is carried out $|$ i has occurred)

In a project with stochastic evolution structure, some events may occur and some activities may be carried out several times during one and the same execution of the project. Let D_{ij}^α be the (nonnegative) duration of the αth execution of activity $<i,j>$ ($\alpha\in\mathbb{N}$). Then the second component F_{ij} of the weight of arc $<i,j>$ is the conditional distribution function of D_{ij}^α given that activity $<i,j>$ is carried out for the αth time:

(1.2.4) $$F_{ij}(t):=\begin{cases} P(D_{ij}^\alpha\leq t\,|<i,j>\text{ is carried out for the }\alpha\text{th time}) & \text{if } t\geq 0 \\ 0 & \text{if } t<0 \end{cases}$$

(1.2.5) The definition of p_{ij} and F_{ij} implies that these quantities are independent of how may times the project event i has occurred or, respectively, activity $<i,j>$ has been carried out before. A more precise formulation of that assumption will be given in section 1.3. It is therefore justified to speak of the **duration of activity** $<i,j>$ denoted by D_{ij}, which corresponds to the duration of any execution of activity $<i,j>$. We assume that for each activity $<i,j>$ the expected duration $\mathbb{E}(D_{ij})$ [1] is finite and thus $P(D_{ij}<\infty)=1$.

If a GERT network has more than one source, a so–called **initial distribution** must be given in addition to the arc weights. Let $\mathbb{P}(M)$ be the power set of M (that is, the set of all subsets of M) and $\mathbb{P}(M):=\mathbb{P}(M)\setminus\{\emptyset\}$ be the set of all nonempty subsets of M and let

(1.2.6) $$q_{R'}:=P[(\mathcal{A}_i)_{i\in R'}\cap(\overline{\mathcal{A}}_k)_{k\in R\setminus R'}]\quad\text{for }R'\in\mathbb{P}(R)$$

be the probability that exactly all sources $i\in R'\subseteq R$ are activated (that is, all sources $i\in R'$ are activated but none of the remaining sources $k\in R\setminus R'$ is activated). In (1.2.6) $(\mathcal{A}_i)_{i\in R'}$ and $(\overline{\mathcal{A}}_k)_{k\in R\setminus R'}$ stand for $\bigcap_{i\in R'}\mathcal{A}_i$ and $\bigcap_{k\in R\setminus R'}\overline{\mathcal{A}}_k$, respectively. If

(1.2.7) $$\sum_{R'\in\mathbb{P}(R)}q_{R'}=1$$

---

[1]   Strictly speaking $\mathbb{E}(D_{ij})$ represents the conditional expectation of the duration of activity $<i,j>$ given that $<i,j>$ is carried out.

that is, the probability that at least one source is activated equals 1, then by

(1.2.8) $q(R) := \{q_{R'} \mid R' \in \mathbb{P}(R)\}$

a discrete probability distribution is given.

Let T_i be the time of activation of node i if node i is activated at most once and where we put $T_i := \infty$ if node i is not activated. Thus,

$$A_i = \{T_i < \infty\}, \quad \overline{A}_i = \{T_i = \infty\}$$

For $R' \in \mathbb{P}(R)$, let $t_{R'}$ be the vector with the components t_i ($i \in R'$, $t_i \in \mathbb{R}$). To have uniqueness, we order the components of $t_{R'}$ according to increasing node numbers. For example, if $R' = \{1,3,4\}$, then $t_{R'} = \begin{bmatrix} t_1 \\ t_3 \\ t_4 \end{bmatrix}$. Furthermore, let

(1.2.9) $\mathcal{H}_{R'}(t_{R'}) := P[(T_i \leq t_i)_{i \in R'} \cap (\overline{A}_k)_{k \in R \setminus R'}]$ for $t_i \in \mathbb{R}$, $i \in R' \in \mathbb{P}(R)$

be the probability that each of the sources $i \in R'$ is activated by time t_i whereas none of the sources $k \in R \setminus R'$ is activated. Then

$$\lim_{t_{R'} \to \infty} \mathcal{H}_{R'}(t_{R'}) = q_{R'}$$

where $t_{R'} \to \infty$ means "$t_i \to \infty$ for all $i \in R'$". For $q_{R'} > 0$, $\dfrac{\mathcal{H}_{R'}}{q_{R'}}$ is a distribution function. If each source of the GERT network in question which is activated is activated at time 0, then for each $R' \in \mathbb{P}(R)$

$$\mathcal{H}_{R'}(t_{R'}) = \begin{cases} q_{R'} & \text{for } t_i \geq 0 \ (i \in R') \\ 0, & \text{otherwise} \end{cases}$$

The family of functions

(1.2.10) $\mathcal{X}(R) := \{\mathcal{X}_{R'} \mid R' \in \mathbb{P}(R)\}$

is called the **initial distribution** of the GERT network in question.

If the GERT network has only one source i, we do not need the initial distribution function $\mathcal{X}_i$ because we will stipulate later that each project execution begins with the activation of at least one source at time 0.

In addition to the probability $q_{R'}$, we need the probability

(1.2.11) $q_{R'} := \begin{cases} P[(\mathcal{A}_i)_{i \in R'}] & \text{for } R' \in \mathbb{P}(R) \\ 1 & \text{for } R' = \emptyset \end{cases}$

that (at least) all sources $i \in R'$ are activated. The probabilities $q_{R'}$ can be expressed in terms of the probabilities $q_{\bar{R}}$ and vice versa $(R', \bar{R} \in \mathbb{P}(R))$ using formulas from elementary probability theory. To express $q_{R'}$ in terms of $q_{\bar{R}}$ with $\bar{R} \in \mathbb{P}(R)$, let A be any random event and $B_1, \ldots, B_r$ be disjoint random events such that $A \subseteq \bigcup_{\rho=1}^{r} B_\rho$. Then it holds that

$$A = \bigcup_{\rho=1}^{r} (A \cap B_\rho)$$

and thus

(1.2.12) $P(A) = \sum_{\rho=1}^{r} P(A \cap B_\rho)$

Now let $R_1, \ldots, R_r$ be the nonempty subsets of R, let B_ρ be the random event "exactly the sources $i \in R_\rho$ are activated" ($\rho = 1, \ldots, r$), and let A be the random event "(at least) the sources $i \in R' \neq \emptyset$ are activated". Taking $\bar{R}$ as a representative of the sets $R_1, \ldots, R_r$, (1.2.11) and (1.2.12) yield

(1.2.13)
$$q_{R'} = \sum_{\bar{R} \in \mathbb{P}(R)} P\{(A_i)_{i \in R'} \cap [(A_k)_{k \in \bar{R}} \cap (\bar{A}_l)_{l \in R \setminus \bar{R}}]\}$$

$$= \sum_{\substack{\bar{R} \in \mathbb{P}(R) \\ \bar{R} \supseteq R'}} P[(A_k)_{k \in \bar{R}} \cap (\bar{A}_l)_{l \in R \setminus \bar{R}}] = \sum_{\substack{\bar{R} \in \mathbb{P}(R) \\ \bar{R} \supseteq R'}} q_{\bar{R}} \quad \text{for } R' \in \mathbb{P}(R)$$

In particular, we have

$$q_{R'} = q_{R'} \quad \text{for } R' = R$$

To express $q_{R'}$ in terms of $q_{\bar{R}}$ with $\bar{R} \in \mathbb{P}(R)$, let A be the random event "(at least) the sources $i \in R'$ are activated", let B be the random event "(at least) one of the sources $k \in R \setminus R'$ is activated", and let $\bar{B}$ be the complement of B with respect to R, that is, the random event "(at least) the sources $k \in R \setminus R'$ are not activated". Applying the relation

$$P(A \cap \bar{B}) = P(A) - P(A \cap B)$$

and the definition of $q_{R'}$ we obtain

(1.2.14)
$$q_{R'} = P[(A_i)_{i \in R'}] - P[(A_i)_{i \in R'} \cap (\bigcup_{k \in R \setminus R'} A_k)]$$

Suppose $R \setminus R' = \{1, \ldots, r\}$ and $B_\rho := (A_i)_{i \in R'} \cap A_\rho$ ($\rho = 1, \ldots, r$). Then $A \cap B = \bigcup_{\rho=1}^{r} B_\rho$ and

$$P[(A_i)_{i \in R'} \cap (\bigcup_{k \in R \setminus R'} A_k)] = P(\bigcup_{\rho=1}^{r} B_\rho)$$

$$= \sum_{\rho=1}^{r} P(B_\rho) - \sum_{\substack{\rho_1, \rho_2 = 1 \\ \rho_1 < \rho_2}}^{r} P(B_{\rho_1} \cap B_{\rho_2}) + - \ldots + (-1)^{r+1} P(\bigcap_{\rho=1}^{r} B_\rho)$$

$$= \sum_{\tilde{R}\in\mathbb{P}(R\backslash R')} (-1)^{|\tilde{R}|+1} P\{\bigcap_{k\in\tilde{R}} [(A_i)_{i\in R'} \cap A_k]\}$$

$$= \sum_{\tilde{R}\in\mathbb{P}(R\backslash R')} (-1)^{|\tilde{R}|+1} P[(A_i)_{i\in\tilde{R}\cup R'}]$$

Inserting this result into (1.2.14) and observing the definition of $q_{R'}$ gives

(1.2.15) $$q_{R'} = q_{R'} + \sum_{\tilde{R}\in\mathbb{P}(R\backslash R')} (-1)^{|\tilde{R}|} q_{\tilde{R}\cup R'}$$

$$= \sum_{\tilde{R}\in\mathbb{P}(R\backslash R')} (-1)^{|\tilde{R}|} q_{\tilde{R}\cup R'} = \sum_{\substack{\overline{R}\in\mathbb{P}(R)\\ \overline{R}\supseteq R'}} (-1)^{|\overline{R}\backslash R'|} q_{\overline{R}} \quad \text{for } R'\in\mathbb{P}(R)$$

Finally we deal with **GERT nodes**. There are six different node types used in GERT networks, which result from combining three different entrance sides and two exit sides of a node, that is, we think of a node as being decomposed into an entrance side and an exit side.

At first we consider three types of **entrance sides** of a node i.

(1.2.16) If, during a project execution, project event i occurs exactly at the time at which *all* activities leading into node i are terminated for the first time, we say that node i has an **AND entrance**, and we use the symbol shown in Fig. 1.2.1. The time of occurrence of event i is thus the latest of the first times of completion of the incoming activities. The AND entrance is the entrance side of nodes used in CPM and PERT networks.

(1.2.17) We speak of an **inclusive–or entrance** or briefly **IOR entrance** (see Fig. 1.2.1) if event i occurs precisely at the time at which the *first* (with respect to time) of the activities leading into node i is terminated for the first time. The time of occurrence of event i is the earliest of the first times of completion of the incoming activities.

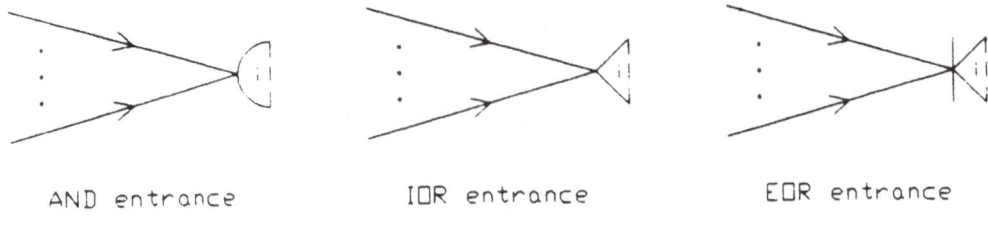

AND entrance   IOR entrance   EOR entrance

Figure 1.2.1

(1.2.18) Node i is said to have an **exclusive–or entrance** or briefly **EOR entrance** (compare Fig. 1.2.1) if event i occurs exactly *every time* at that point in time at which one of the activities leading into node i is terminated.

(1.2.19) Note that nodes with AND entrance or IOR entrance can, according to definition, be activated at most once during one project realization, whereas nodes with EOR entrance may be activated several times during a single project realization. If $r > 1$ completions of activities leading into a node with EOR entrance occur at one and the same point in time t, then, by definition, the node is activated r times at time t. Later we will state some assumptions which ensure that different activations of a node with EOR entrance occur one after another. Then in definition (1.2.18) of the EOR entrance, "at which one of the activities" can be replaced by "at which exactly one of the activities".

We now consider two types of **exit sides** of a node i.

(1.2.20) If *all activities* emanating from node i are carried out when project event i has occurred during a project execution, we say that node i has a **deterministic exit** (see Fig. 1.2.2). This is the exit side of nodes used in CPM and PERT networks.

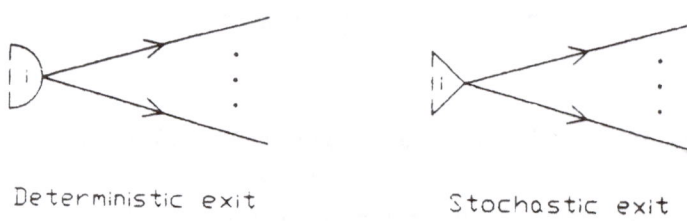

Deterministic exit   Stochastic exit

Figure 1.2.2

(1.2.21) If *exactly one* of the activities emanating from node i is carried out when project event i has occurred, we speak of a **stochastic exit** (compare Fig. 1.2.2).

If node i has a deterministic exit, it holds that

$$p_{ij}=1 \text{ for all } j \in \mathcal{S}(i)$$

and for a node i with stochastic exit we have

$$\sum_{j \in \mathcal{S}(i)} p_{ij}=1$$

(1.2.22) Note that (1.2.20) and (1.2.21) do not imply that those outgoing activities which are carried out after the activation of a node are begun at the time of occurrence of their initial event (that is, at their earliest possible start time). In time planning of projects, however, where we are interested in completing a project as early as possible, we stipulate that *each activity is begun at its earliest possible start time*. In chapter 5 we will see that in project scheduling with precedence constraints given by a GERT network, the execution of some activities may be delayed due to limited capacity of resource required.

From the definition of the AND entrance and IOR entrance we see that nodes with that entrance side are of any interest only if there are at least two incoming arcs. Similarly, it only makes sense to assign the deterministic exit to a node if there are at least two outgoing arcs. Hence, we establish the following

(1.2.23)   Convention.
Each node i with $|\mathcal{P}(i)| \leq 1$ has an EOR entrance and each node i with $|\mathcal{S}(i)| \leq 1$ has a stochastic exit.

PERT node   STEOR node

Figure 1.2.3

(1.2.24) As already stated, in CPM and PERT networks each node has an AND
entrance and a deterministic exit (the symbol used for such a "PERT node" is shown in
Fig. 1.2.3). That node type of a GERT network which is most easily handled (as we
will see later) is the node with EOR entrance and stochastic exit called **STEOR node**
(the symbol is also shown in Fig. 1.2.3). The remaining five node types are referred to
as **non–STEOR nodes**. A STEOR node represents some kind of **transit node** or **through
node**: Whenever one incoming activity has been completed, one outgoing activity is
carried out. A GERT network all of whose nodes are STEOR nodes is called a **STEOR
network**. We speak of an **AND node** (or **IOR node** or **EOR node**, respectively) if the
node has an AND entrance (or IOR entrance or EOR entrance, respectively). Similarly,
a node with a deterministic exit (or stochastic exit, respectively) is called a
deterministic node (or **stochastic node**, respectively).

In summary, we have the following

(1.2.25)   Definition.

A GERT network is a project network with sources and sinks based on
activity–on–arc representation where each node belongs to one of the six
node types resulting from combination of the three entrance sides and two
exit sides introduced above, where a weight vector $\begin{bmatrix} p_{ij} \\ F_{ij} \end{bmatrix}$ with the
aforementioned meaning is assigned to each arc $<i,j>$, and where an initial
distribution of the network is given.

At this point we return to the example of the development of a new product sketched at
the beginning of section 1.2. It deals with the development of one product each by two
different research teams A and B. As a rule, if the test of a product after its development
proves the product to be unusable, there follow a product improvement and a new
product test. The work is discontinued only if the product is completely unsuccessful.
The entire development project is successfully terminated if at least one of the two
teams has developed a usable new product. The associated GERT network N (without
arc weights) is shown in Fig. 1.2.4. N has the source 1 and the sinks 6 and 7.
Furthermore, N contains cycles, which represent an essential new element of GERT
networks as compared to the "classical" CPM and PERT networks.

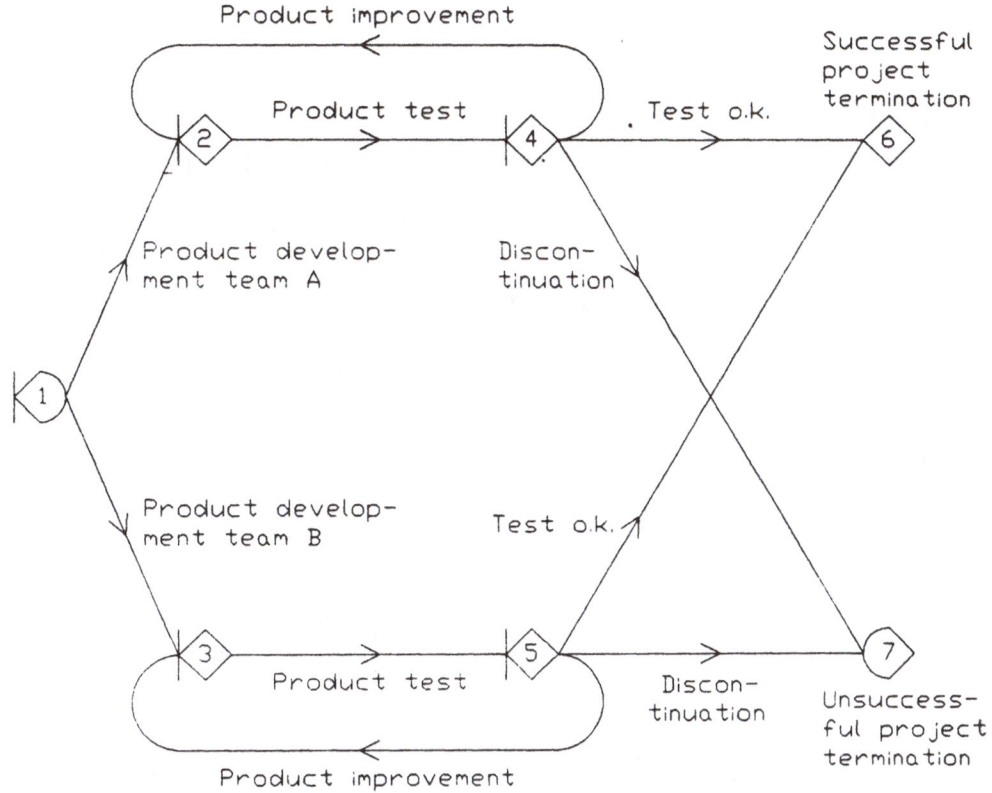

Figure 1.2.4

1.3 Assumptions and Structural Problems

In this section we state several assumptions which are required for GERT networks, and we discuss some structural properties of GERT networks.

The first assumption is concerned with the case of several sources. It says that, figuratively speaking, those parts of the project in question whose executions are induced by distinct beginning events do not interfere, more precisely:

Assumption A1.

During each project execution, at most one of those sources is activated from which one and the same sink is reachable.

Moreover, we establish the

(1.3.1) **Convention.**
Each project execution begins with the activation of at least one source at time 0.

Convention (1.3.1) implies that

$$\mathcal{X}_{R'}(t_{R'}) = 0 \quad \text{if } t_i < 0 \text{ for some } i \in R' \quad (R' \in \mathbb{P}(R))$$

and

$$\sum_{R' \in \mathbb{P}(R)} \mathcal{X}_{R'}(0) = 1$$

In the latter formula, 0 means the null vector with $|R'|$ components.

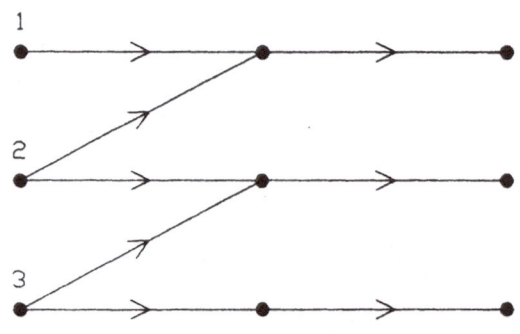

Figure 1.3.1

Assumption A1 says for the GERT network in Fig. 1.3.1 (where the weights are omitted and the node types are not specified) that both the sources 1 and 3 may be activated during one and the same network realization but not the sources 1 and 2 or the sources 2 and 3.

(1.3.2)   Remarks.

(a) $R' \in \mathbb{P}(R)$ is called a **feasible subset** of R if $\mathcal{Z}(i) \cap \mathcal{Z}(j) = \emptyset$ for all $i, j \in R'$ with $i \neq j$. In particular, all the singletons $\{i\}$ with $i \in R$ are feasible subsets of R, and if $R_1 \in \mathbb{P}(R)$ is not feasible, then each $R_2 \supseteq R_1$ $(R_2 \in \mathbb{P}(R))$ is not feasible, either. The set of all feasible subsets of R is denoted by $\hat{\mathbb{P}}(R)$. For the network in Fig. 1.3.1, $\hat{\mathbb{P}}(R)$ consists of the elements $\{1\}$, $\{2\}$, $\{3\}$, and $\{1,3\}$.

(b) Because of assumption A1 we have

(1.3.3) $q_{R'} = 0$ for $R' \in \mathbb{P}(R) \setminus \hat{\mathbb{P}}(R)$

and thus by (1.2.7)

$$\sum_{R' \in \hat{\mathbb{P}}(R)} q_{R'} = 1$$

Moreover, by (1.2.13)

$$q_{R'} = \sum_{\substack{\bar{R} \in \hat{\mathbb{P}}(R) \\ \bar{R} \supseteq R'}} q_{\bar{R}} \quad \text{for } R' \in \mathbb{P}(R)$$

and in particular

$$q_{R'} = 0 \quad \text{for } R' \in \mathbb{P}(R) \setminus \hat{\mathbb{P}}(R)$$

Hence, only feasible subsets of R need to be considered in what follows.

(c) Instead of assumption A1 it suffices to require that (1.3.3) holds to be true, that is, A1 is satisfied with probability 1.

(1.3.4) Next we show that each GERT network N with several sources can be transformed into a GERT network with only one source, which is called the **one–source network corresponding to** N. To do so we introduce a STEOR node i_0 (the new single source) and, for each $R' \in \hat{\mathbb{P}}(R)$, an EOR node $i_{R'}$, where $i_{R'}$ has a stochastic exit if

$|R'| = 1$ and a deterministic exit if $|R'| > 1$. Moreover, for each $R' \in \hat{P}(R)$, we introduce the arc $\langle i_0, i_{R'} \rangle$ and $|R'|$ dummy activities $\langle i_{R'}, i \rangle$ $(i \in R')$. The weight of arc $\langle i_0, i_{R'} \rangle$ is $\begin{bmatrix} q_{R'} \\ \pi_{R'}/q_{R'} \end{bmatrix}$ for $q_{R'} > 0$ and $\begin{bmatrix} 0 \\ F \end{bmatrix}$ for $q_{R'} = 0$ where F may be any distribution function, for example, the unit step function with jump point at 0 (that is, $\langle i_0, i_{R'} \rangle$ represents a dummy activity). Note that the original sources $i \in R$ keep their EOR entrance.

As an example, we consider the network in Fig. 1.3.1. The corresponding one–source network (without arc weights) is shown in Fig. 1.3.2, where dummy activities are indicated by dashed–line arrows.

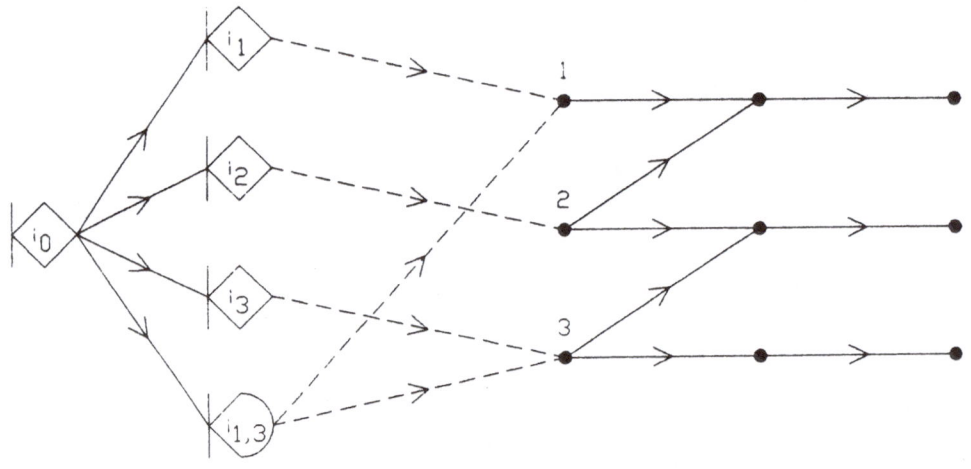

Figure 1.3.2

Suppose that exactly the sources $i \in R'$ are activated during a realization of a GERT network N. Then assumption A1 implies that in the corresponding one–source network the source i_0 is activated at time 0 and $|R'|$ different walks are executed, which are disjoint aside from the common arc $\langle i_0, i_{R'} \rangle$ and the nodes i_0 and $i_{R'}$.

In section 2.2 we will see that a GERT network with several sources is "equivalent" to its corresponding one–source network as far as the temporal analysis of GERT networks is concerned. The transformation of a GERT network into the corresponding one–source network, however, may cause additional computational effort or the network may loose some nice properties through that transformation (for example, the one–source network corresponding to a STEOR network with several sources does not represent a STEOR network unless all feasible subsets $R' \in \hat{P}(R)$ are one–element sets). Therefore, we will generally deal with GERT networks with several sources in chapters 1 to 4 unless it is stated explicitly that the network in question has only one source.

The second assumption A2, which expresses some independence properties, is relatively complicated due to the complex structure of GERT networks. Thus, we only give an illustrative formulation of A2, which is sufficient for what follows and consists of four parts (a) to (d). After that we present some interpretations and consequences of assumption A2. A precise formulation of A2 requires a detailed look at the stochastic processes related to GERT networks (cf. Neumann (1984b)) which is beyond the scope of this monograph. In chapter 3, we will discuss only the stochastic processes associated with STEOR networks. To formulate assumption A2 we introduce some terminology and notation.

For $i \in R$, let $N(i)$ be the subnetwork of the underlying GERT network N which is induced by $\mathcal{R}(i)$ and where the nodes belonging to $N(i)$ are supposed to be of the same type as the corresponding nodes in N except that convention (1.2.23) has to be satisfied.

An **activity sequence** corresponds to a walk in N. A set $\mathcal{W}$ of **activity sequences** is called feasible if for all $W_1, W_2 \in \mathcal{W}$ with $W_1 \neq W_2$, the two walks corresponding to W_1 and W_2 are disjoint except that they share the same deterministic initial node and may share the same final node. For example, in Fig. 1.3.3 the activity sequences $\{<1,2>,<2,4>\}$, $\{<1,3>,<3,4>\}$ and $\{<1,5>\}$ on the one hand and $\{<1,2>,<2,4>\}$, $\{<1,3>,<3,5>\}$ and $\{<1,5>\}$ on the other hand constitute two feasible sets of activity sequences. Two different elements of a feasible set of activity sequences correspond to partial projects that can be executed simultaneously in analogy to two different sources from a feasible subset of R that can be activated simultaneously.

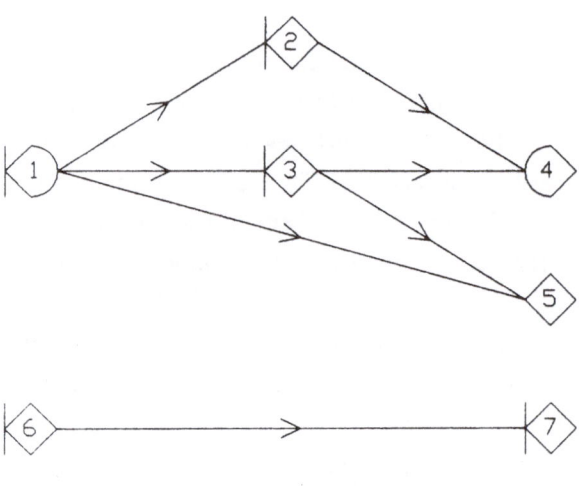

Figure 1.3.3

Let D_{ij}^{α} again denote the duration of the αth execution of activity $<i,j>$ ($\alpha\in\mathbb{N}$). For a stochastic node i, let B_i^{β} be that activity with initial node i which is carried out when i has been activated for the βth time ($\beta\in\mathbb{N}$).

(1.3.5) Let the **state of the project at time** $t\geq0$ be given by the set of all nodes activated at time t and all activities being carried out at time t. The **past history up to time** t comprises all activations of nodes occurred at time $<t$ (including random variable B_i^{β} if node i is stochastic and its βth activation occurred at time $<t$) and all executions of activities terminated at time $<t$ (including random variable D_{ij}^{α} if the αth execution of activity $<i,j>$ was terminated at time $<t$). The **project evolution beginning at time** t comprises all activations of nodes occurring at time $\geq t$ and all executions of activities beginning at time $\geq t$.

Then assumption A2 reads as follows.

Assumption A2.
   (a) For each feasible subset R' of R and for all $i,j\in R'$ with $i\neq j$, the evolution of the partial project corresponding to $N(i)$ does not have any influence on the evolution of the partial project corresponding to $N(j)$, in short, $N(i)$ and $N(j)$ are independent.

(b) For each feasible set $\mathcal{W}$ of activity sequences and for all $W_1, W_2 \in \mathcal{W}$ with $W_1 \neq W_2$, the evolution of the partial project corresponding to W_1 does not influence the evolution of the partial project corresponding to W_2, in short, W_1 and W_2 are independent.

(c) For each activity $<i,j>$ and for all $\alpha, \beta \in \mathbb{N}$ and $t \geq 0$, the probability

$$P(D_{ij}^{\alpha} \leq t \mid <i,j> \text{ is carried out for the } \alpha \text{th time})$$

is independent of α and, if i is a stochastic node, the probability

$$P(B_i^{\beta} = <i,j> \mid i \text{ has been activated for the } \beta \text{th time})$$

is independent of β.

(d) For every $t \geq 0$, the project evolution beginning at time t is conditionally independent of the past history up to time t provided that the state of the project at time t is known [2].

(1.3.6) **Remarks.**

(a) The **independence assumption** A2a is an addition to assumption A1. These two assumptions imply that for any feasible subset R' of R

$$q_{R'} = \prod_{i \in R'} q_i$$

and for every $S' \in \mathbb{P}(S), S'(i) := S' \cap \mathcal{R}(i)$ and all $t_j, \tau_i \in \mathbb{R}_+$

$$P\left[\bigcap_{l \in R'} \bigcap_{j \in S'(i)} (T_j \leq t_j) \mid (T_i = \tau_i)_{i \in R'} \cap (\mathcal{A}_k)_{k \in R \backslash R'}\right]$$
$$= \prod_{i \in R'} P\left[(T_j \leq t_j)_{j \in S'(i)} \mid T_i = \tau_i, (\mathcal{A}_k)_{k \in R \backslash \{i\}}\right]$$

---

[2] For the concept of conditional independence see Cinlar (1975), section 2.2.

(b) Assumption A2a expresses an **independence property** similar to assumption A2b in
the following sense: If we additionally define a set of activity sequences $\mathcal{W}$ to be
feasible when for all $W_1, W_2 \in \mathcal{W}$ with $W_1 \neq W_2$, the initial nodes of W_1 and W_2 are distinct
sources that belong to one and the same feasible subset R' of R, then A2a implies
that for each feasible $\mathcal{W}$ and all $W_1, W_2 \in \mathcal{W}$ with $W_1 \neq W_2$, W_1 and W_2 are independent.

For example, the activity sequences $\{<1,2>,<2,4>\}$ and $\{<6,7>\}$ in Fig. 1.3.3 form
one feasible set of activity sequences, where the activity sequences are independent.
Assumptions A2a and A2b imply that the durations of executions of activities that
belong to distinct activity sequences from one and the same feasible set of activity
sequences are independent. Analogously, for fixed $\beta_i, \beta_j \in \mathbb{N}$, the random variables
$B_i^{\beta_i}$ and $B_j^{\beta_j}$ where $i \neq j$ and i and j belong to distinct activity sequences from one
and the same feasible set of activity sequences are independent.

(c) Assumption A2c represents some **homogeneity property** and says in particular that
the arc weights p_{ij} and F_{ij} are independent of how many times project event i
has occurred or, respectively, activity $<i,j>$ has been carried out before. It is
therefore justified to speak of the **duration** D_{ij} **of activity** $<i,j>$, which
corresponds to the duration of any execution of $<i,j>$ (cf. (1.2.5)). Let the two
random events "(activity) $<i,j>$ (is) carried out" and "(node) i (has been)
activated" be such that the execution of activity $<i,j>$ is induced by the
activation of node i (that is, there is no activity execution between the activation
of i and the execution of $<i,j>$). Then relation $P(A \cap B | C) = P(A | B \cap C) P(B | C)$
provides

$$P(<i,j> \text{ carried out}, D_{ij} \leq t \,|\, i \text{ activated})$$
$$= P(<i,j> \text{ carried out} \,|\, i \text{ activated}) P(D_{ij} \leq t \,|\, <i,j> \text{ carried out})$$
$$= p_{ij} F_{ij}(t)$$

for all $t \geq 0$. Note that in our case $B \subseteq C$ and thus $B \cap C = B$.

(d) Assumption A2d expresses the following **Markov property**: As far as predicting the
project evolution beginning at time t is concerned, all information on the past
history becomes worthless once the project state at time t is given. Moreover, since

the random variables D_{ij}^{α} and B_i^{β} are independent of the point in time at which project event i has occurred, the project evolution beginning at time t depends only on the project state at time t and not on the point in time t itself (in other words, we have some independence of translations of the initial moment on the time axis). This means that the stochastic process which describes the evolution in time of the project is **stationary**.

(e) It is easy to see that if a GERT network N with several sources satisfies assumptions A1 and A2a to A2d, then the one–source network corresponding to N fulfills assumptions A2b, A2c and A2d.

The next assumption represents an additional requirement for the arc weights of a GERT network, which is generally satisfied in applications.

Assumption A3.

For each node k of a cycle structure C, there is a path from k to a node outside C such that $p_{ij} > 0$ for every arc $\langle i, j \rangle$ of this path.

Assumption A3 says, figuratively speaking, that each cycle structure is "left" with positive probability. A3 is satisfied in particular if all activities of the network have positive execution probability.

(1.3.7)   Remark.

A1, A2, and A3 are basic assumptions supposed to be satisfied by each GERT network.

In real projects, we generally have that each activity outside any cycle (from which there is thus no return to events that have already occurred once before) can be carried out at most once during a single project realization, whereas activities within cycles may be carried out several times. This property is utilized in the temporal analysis of GERT networks. To ensure that a GERT network has that property, some additional assumptions are stated.

Assumption A4.

Every node belonging to a cycle is a STEOR node.

Assumption A5.

Every node with at least two predecessors that does not belong to any cycle has an AND entrance or IOR entrance.

Assumption A6.

During each realization of a GERT network N, at most one entrance arc of each cycle structure of N is executed.

(1.3.8)   Definition.

A GERT network is called **weakly admissible** if it satisfies assumption A4 and **admissible** if it satisfies the assumptions A4, A5, and A6 (in addition to A1, A2, and A3).

Since STEOR nodes behave like "transit nodes", assumption A4 implies that nodes and arcs within cycles can be activated or, respectively, executed several times during a single project realization. From A5 together with the definition of AND nodes and IOR nodes it follows that all nodes before any cycle structure of rank 1 (cf. (1.1.2)) are activated at most once during a single project realization. This fact together with assumptions A4 and A6 and the "transit property" of STEOR nodes ensure that the nodes of each cycle structure of rank 1 are activated "one after another" [3] and that at most one exit arc of each such cycle structure is executed, and this exit arc is carried out only once. By the same reasoning we see that the latter fact holds for all cycle structures and that all nodes and arcs outside cycles are activated or, respectively, executed at most once. In particular, each terminal event of the project occurs at most once. Since all activity durations are finite with probability 1 (cf. (1.2.5)), the total time of carrying out all activities outside cycles during one project execution is finite with probability 1. In section 3.2 we will see that assumptions A3 and A4 ensure that, with probability 1, there is a finite number of node activations in each cycle structure C and a finite time between the moment of activation of one of the entrance nodes of C and the moment at which C is subsequently "left" (cf. corollary (3.2.9)). Hence, the total time of executing all activities that are carried out during the project execution in question is finite with probability 1. In summary, we have

---

[3]   In particular, different activations of one and the same node in a cycle occur one after another (compare (1.2.19)).

(1.3.9)   Theorem.

For an admissible GERT network, all nodes and arcs outside any cycle are activated or, respectively, executed at most once during one project execution. The number of all node activations and activity executions during a single project execution is finite with probability 1. Moreover, the total time of executing all activities that are carried out during a single project execution is finite with probability 1.

It is rather simple to test whether assumptions A3, A4, and A5 are satisfied in a given GERT network (this can be done in $O(|E|)$ time). On the other hand, it has been shown that even for acyclic GERT networks with only one source and only one sink, the decision problem whether there is a network realization in which the activation of the source implies the activation of the sink is NP–complete (that is, presumably not solvable in polynomial time), aside from PERT networks and from GERT networks without AND nodes (cf. Siedersleben (1981)). As a consequence, the problem of testing whether assumption A6 is satisfied for a given GERT network is in general NP–hard. The testing of assumption A1 or condition (1.3.3), respectively, cannot be done in polynomial time, either, because all nonempty subsets of R have to be examined. However, since the number of sources is generally very small in practice, the testing of assumption A1, as a rule, does not take much time. Assumption A2 stipulates the independence of the random variables that describe the stochastic behaviour of the project in question and is supposed to be satisfied a priori.

(1.3.10) Let N be a weakly admissible GERT network and i be an IOR node of N. If at most one arc leading into node i is executed during each realization of N and if that arc is executed only once, then i is called a **non–genuine IOR node** because it behaves like an EOR node. The problem of finding all non–genuine IOR nodes in a GERT network is, in general, NP–hard. An algorithm that discovers all non–genuine IOR nodes and tests whether assumption A6 is satisfied can be found in Neumann and Steinhardt (1979a), section 4.4. For the so–called BES networks, which will be discussed in chapter 4, the non–genuine IOR nodes can be determined in the course of the BES method (the procedure for evaluating BES networks), see section 4.6.

Since nodes with EOR entrance can most easily be handled (as we will see later), we establish the following

(1.3.11)   Convention.

Each non–genuine IOR node of a weakly admissible GERT network is
replaced by an EOR node (with the same exit side).

Note that convention (1.3.11) means some modification of assumption A5, which is
supposed to be made in what follows. It can immediately be seen that in a weakly
admissible GERT network without deterministic nodes, all IOR nodes are non–genuine
and thus have to be replaced by EOR nodes. If, in addition, the network does not
contain AND nodes, we obtain a STEOR network.

(1.3.12)   Remark.

If a GERT network is weakly admissible or admissible, the corresponding one–source
network is also weakly admissible or admissible, respectively, and meets conventions
(1.2.23) and (1.3.11).

(1.3.13) At last we make precise the concept of the project duration for GERT
networks. In classical project planning, as long as activities of the project are ready for
execution (that is, when their initial events have occurred), they are really carried out.
For GERT networks, it also makes sense to consider the case where the execution of
activities is discontinued when no more sink can be activated (this is called the **case of
skipping**). Let D^{skip} denote the **skipping project duration**, that is, the time elapsed
between the beginning of the project and the earliest point in time after which no more
sink is activated, where we put $D^{skip}(\omega) := \infty$ if no sink is activated during project
realization ω. The nonskipping project duration or briefly **project duration**, denoted by
D, is the time elapsed between the beginning of the project and the earliest point in time
after which no more activities are being carried out. Sometimes, the skipping project
duration D^{skip} is of greater importance in practice than the nonskipping project
duration D, namely if there is not much sense in carrying out activities after the
occurrence of the last realizable terminal event of the project.

(1.3.14) For admissible GERT networks it holds that $P(D<\infty)=1$ (compare theorems
(1.3.9) and (3.2.10)). Moreover, we have $D^{skip}=D$ for PERT networks and for STEOR
networks with only one source (cf. theorem (3.2.6)). In general, it holds that $D^{skip} \leq D$
provided that at least one sink is activated during each project realization. Trivially,
$D(\omega) \leq D^{skip}(\omega)$ if no sink is activated during project realization ω.

1.4 Complete and GERT Subnetworks

Let N' be a subnetwork of a GERT network N (compare (1.1.5)) with node set V' and source set R', where the nodes belonging to N' are supposed to be of the same type as the corresponding nodes in N. In general N' does not represent a GERT network when treated as a separate entity even if N' is induced by V'. For example, if some successors of a stochastic node i of N' are in N' and some are not in N' (thus, node i is an exit node but not a sink of N') and if an exit arc of N' with initial node i has positive execution probability, then the sum of the execution probabilities of the arcs that emanate from i and belong to N' is less then 1 (compare Fig. 1.4.1 where the arcs emanating from node i are marked with their execution probabilities). Thus, node i is neither a deterministic node nor a stochastic node "within N'".

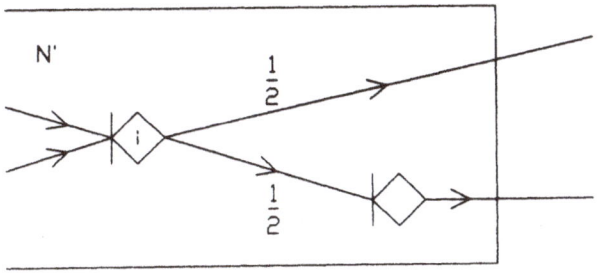

Figure 1.4.1

Furthermore, in a subnetwork N' with $|R'|>1$, assumption A1 or condition (1.3.3), respectively, must hold if N' taken on its own is to represent a GERT network. The initial distribution of N', however, need not be specified since we will later evaluate only subnetworks that can be reduced to STEOR networks, where we merely consider the transitions from their sources to their sinks.

To eliminate the aforementioned shortcomings we first establish the following

(1.4.1)   **Convention.**
           Each entrance node of a subnetwork N' is a source of N', and each exit node of
           N' is a sink of N'.

Convention (1.4.1) ensures that N' represents a network with sources and sinks if taken on its own (cf. remark (1.1.9c)) and that all entrance nodes and exit nodes of N' are of the same type in N and in N'. Furthermore, convention (1.4.1) implies that no entrance node and no exit node of N' belong to a cycle which lies entirely in N'. However, there may be an entrance node or an exit node of N' that belong to a cycle C in N where C does not lie entirely in N'.

If we take a subnetwork N' from a given GERT network N, convention (1.4.1) can always be satisfied, if necessary, by adding dummy activities (and auxiliary nodes) as it is shown in Fig. 1.4.2 where dotted–line arrows represent walks and dashed–line arrows again represent dummy activities.

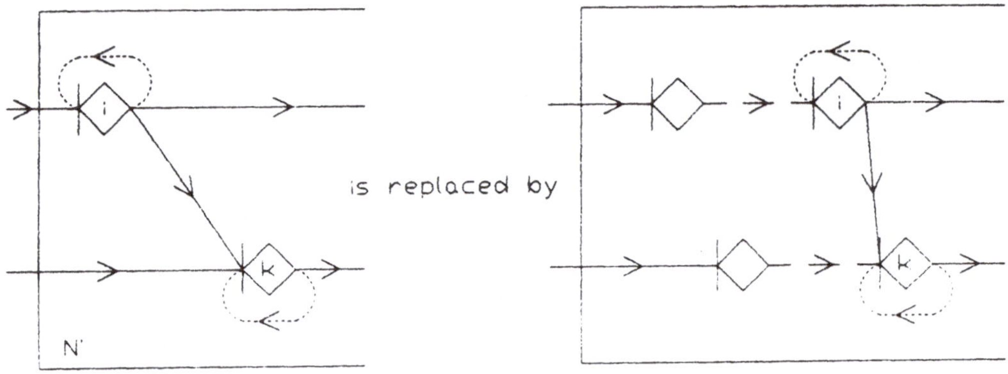

Figure 1.4.2

To meet condition (1.2.23) if a subnetwork N' is considered a separate entity, we establish

(1.4.2)   Convention.

Each entrance node of a subnetwork N' has an EOR entrance, and each exit node has a stochastic exit.

Figure 1.4.3

Fig. 1.4.3 illustrates that convention (1.4.2) can also be satisfied, if necessary, by adding dummy activities. Note that not only the subnetwork N' in Fig. 1.4.3 has to be modified but also the GERT network N. In section 2.2 we will see that the modifications of a weakly admissible GERT network N and a subnetwork N' of N as illustrated in Figs. 1.4.2 and 1.4.3 to satisfy conventions (1.4.1) and (1.4.2) do not change the quantities that provide the temporal analysis of N (compare remark (2.2.18d)). The foregoing observations give rise to the following

(1.4.3)   **Convention and Definition.**
          The nodes of a **subnetwork** N' of a GERT network N are supposed to be of the same type as the corresponding nodes in N. A subnetwork N' of a GERT network N is called **complete** if it is induced by some subset of the node set of N, if it satisfies the conventions (1.4.1) and (1.4.2) and if, aside from convention (1.4.2), the nodes belonging to N' are of the same type as the corresponding nodes in N.

In order that a complete subnetwork N' represents a GERT network if taken on its own, an additional condition has to be satisfied which guarantees that assumption A1 is fulfilled for N'. Thus, we define

(1.4.4)   **Definition.**
          A complete subnetwork N' of a GERT network N is said to be a **GERT subnetwork** if the following condition is met:

          (a)   During each execution of N, at most one of those sources of N' is activated from which one and the same sink of N' is reachable.

Trivially, each complete subnetwork with only one source represents a GERT subnetwork. Condition (1.4.4a) corresponds to assumption A1' and represents a "global" condition whose satisfaction depends not only on the structure of the subnetwork N' but also on that part of the whole GERT network N which lies "before" N' (that is, on the subnetwork of N induced by $\mathcal{R}(R')\backslash R'$).

(1.4.5) An example of a GERT subnetwork is the subnetwork $N(i)$ induced by $\mathcal{R}(i)$ where i is any node of the underlying GERT network N provided that conventions (1.4.1) and (1.4.2) are satisfied. The latter condition is always supposed to be fulfilled if we speak of "the GERT subnetwork $N(i)$" in what follows.

The next theorem is immediate from the definition of a GERT subnetwork.

(1.4.6)   Theorem.
A GERT subnetwork N' of a GERT network represents a GERT network (aside from the initial distribution) if N' is treated as a separate entity.

(1.4.7)   Remarks.
(a) By the **execution** (or, respectively, **realization**) of a GERT subnetwork N' we mean the execution (or, respectively, realization) of N' considered as a separate GERT network.
(b) A GERT subnetwork N' is called **admissible** if N' treated as a separate entity is an admissible GERT network. Trivially, each GERT subnetwork of an admissible GERT network is admissible.

Chapter 2 Temporal Analysis of GERT Networks

In CPM and PERT network techniques, the temporal analysis includes the determination of the earliest and latest times of occurrence of the individual project events besides the computation of the (distribution of the) project duration. For GERT networks, it is also possible to introduce the concepts of earliest and latest times of activation of nodes. Their meaning, however, is different from that for CPM and PERT networks (because project events may occur several times) and their computation is in general much more complicated (cf. Delivorias (1979a, 1979b), Neumann and Steinhardt (1979a), section 2.5, and Wietek (1983)). Therefore, we will not deal with those concepts in this monograph.

Instead, of special importance are quantities connected with the terminal events of the project in question such as the probability that certain terminal events will occur (recall that a GERT network generally has more than one sink or terminal event, respectively) and the conditional distribution function of the (again shortest) project duration or skipping project duration, respectively, given that certain terminal events have occurred. As we will see in section 2.3, those quantities can be computed from the so-called activation functions or the activation distributions of the sinks of the underlying GERT network. The concepts of an activation function and an activation distribution are introduced in section 2.1. In section 2.2 we make precise what is to be understood by the evaluation of an admissible GERT network in the sense of time planning. In section 2.4 we mention some methods for evaluating admissible GERT networks. The temporal analysis of special classes of GERT networks (STEOR networks, EOR networks and BES networks) will be dealt with in chapters 3 and 4.

In time planning of projects modelled by GERT networks, we are interested in terminating the project as early as possible. Thus, we stipulate in what follows that *as soon as* a node i of the respective GERT network has been activated, all outgoing activities are begun if node i is deterministic or exactly one outgoing activity is begun if node i is stochastic. In other words, *each activity is begun at its earliest possible start time* (compare (1.2.22)).

2.1 Activation Functions and Activation Distributions

(2.1.1) Consider an admissible GERT network and let $K_j(t)$ be the number of activations of node j during a project execution in the time interval $[0,t]$. Recall that each project execution begins at time 0 (cf. convention (1.3.1)) and that some nodes may be activated several times during a single project execution.

For $U \subseteq V$, $U \neq \emptyset$ let t_U be the vector with the components t_j ($j \in U$). To have uniqueness we again order the components of t_U according to increasing node numbers. By

$$(2.1.2) \qquad Y_U(t_U) := \begin{cases} \mathbb{E}[\prod\limits_{j \in U} K_j(t_j)] & \text{for } t_j \in \mathbb{R}_+, \ j \in U \\ 0, \ \text{otherwise} \end{cases}$$

where $\mathbb{E}[K]$ is again the expected value of random variable K, a mapping $Y_U : \mathbb{R}^{|U|} \to \mathbb{R}_+$ is given, which is called the **activation function of set** U. The set of functions

$$(2.1.3) \qquad Y(U) := \{Y_{U'} \mid U' \in \mathbb{P}(U)\}$$

is called the **family of activation functions for the set** U.

(2.1.4) If a subscript represents a one–element set, say $U = \{j\}$, we omit the braces in what follows. For example, we write Y_j instead of $Y_{\{j\}}$, and Y_j is referred to as the **activation function of node** j. $Y_j(t)$ is the expected number of activations of node j up to (and including) time t. The quantity

$$z_j := \lim_{t \to \infty} Y_j(t)$$

is called the **activation number of node** j. It represents the expected number of activations of node j during a single project execution. In section 3.4 we will see that z_j exists (and is finite) for each node j of an admissible GERT network.

Now suppose that node j does not belong to any cycle. Since the underlying GERT network is assumed to be admissible, node j is activated at most once during one project execution (cf. theorem (1.3.9)). Let T_j be again the time of activation of node j. Then

$$
(2.1.5) \qquad
\begin{cases}
Y_j(t) = P(T_j \leq t) \quad \text{for } t \geq 0 \\[2mm]
z_j = q_j := P(\mathcal{A}_j) = \lim_{t \to \infty} P(T_j \leq t)
\end{cases}
$$

and, more generally, if none of the nodes of set U belongs to a cycle,

$$
(2.1.6) \qquad Y_U(t_U) = P[(T_j \leq t_j)_{j \in U}]
$$

Let us again assume that none of the nodes of $U \in \mathbb{P}(V)$ belongs to a cycle and let $U' \in \mathbb{P}(U)$. Then

$$
(2.1.7) \qquad \mathcal{Y}_{U'}^{U}(t_{U'}) := P[(T_j \leq t_j)_{j \in U'} \cap (\mathcal{A}_i)_{i \in U \setminus U'}] \quad \text{for } t_j \in \mathbb{R}, j \in U'
$$

is the probability that each of the nodes $j \in U'$ is activated by time t_j whereas none of the nodes $i \in U \setminus U'$ is activated. The corresponding mapping $\mathcal{Y}_{U'}^{U} : \mathbb{R}^{|U'|} \to [0,1]$ is called the **activation distribution of set U' (with respect to set U)** and the set of functions

$$
(2.1.8) \qquad \mathcal{Y}(U) := \{\mathcal{Y}_{U'}^{U} \mid U' \in \mathbb{P}(U)\}
$$

is said to be the **family of activation distributions for the set U.**

The elements of the family $\mathcal{Y}(U)$ can be expressed in terms of elements of the family $Y(U)$ and vice versa using formulas from elementary probability theory. First, we show that $Y_{U'}$ for any $U' \in \mathbb{P}(U)$ can be expressed in terms of $\mathcal{Y}_{\bar{U}}^{U}$ with $\bar{U} \in \mathbb{P}(U)$. Recall the relation

$$
(2.1.9) \qquad P(A) = \sum_{\rho=1}^{r} P(A \cap B_\rho)
$$

where $B_1, \ldots, B_r$ are disjoint random events and A is a random event such that $A \subseteq \bigcup_{\rho=1}^{r} B_\rho$. Now let $U_1, \ldots, U_r$ be the nonempty subsets of U, let B_ρ be the random event "exactly the nodes $k \in U_\rho$ are activated" ($\rho = 1, \ldots, r$), and let A be the random event "(at least) the nodes $j \in U'$ are activated specifically not later than at time t_j". In analogy to the derivation of relation (1.2.13), formula (2.1.9) then provides

$$(2.1.10) \qquad Y_{U'}(t_{U'}) = P[(T_j \leq t_j)_{j \in U'}]$$

$$= \sum_{\bar{U} \in \mathbb{P}(U)} P\{(T_j \leq t_j)_{j \in U'} \cap [(A_k)_{k \in \bar{U}} \cap (\bar{A}_l)_{l \in U \setminus \bar{U}}]\}$$

$$= \sum_{\substack{\bar{U} \in \mathbb{P}(U) \\ \bar{U} \supseteq U'}} \lim_{t_{\bar{U} \setminus U'} \to \infty} P\{(T_j \leq t_j)_{j \in U'} \cap [(T_k \leq t_k)_{k \in \bar{U} \setminus U'} \cap (\bar{A}_l)_{l \in U \setminus \bar{U}}]\}$$

$$= \sum_{\substack{\bar{U} \in \mathbb{P}(U) \\ \bar{U} \supseteq U'}} \lim_{t_{\bar{U} \setminus U'} \to \infty} P[(T_k \leq t_k)_{k \in \bar{U}} \cap (\bar{A}_l)_{l \in U \setminus \bar{U}}]$$

$$= \sum_{\substack{\bar{U} \in \mathbb{P}(U) \\ \bar{U} \supseteq U'}} \lim_{t_{\bar{U} \setminus U'} \to \infty} y_{\bar{U}}^{U}(t_{\bar{U}})$$

where "$t_{\bar{U} \setminus U'} \to \infty$" again means "$t_j \to \infty$ for all $j \in \bar{U} \setminus U'$".

To express $y_{U'}^{U}(t_{U'})$ in terms of $Y_{\bar{U}}(t_{\bar{U}})$ with $\bar{U} \in \mathbb{P}(U)$ we may proceed similarly to the derivation of (1.2.15). We obtain

$$(2.1.11) \qquad y_{U'}^{U}(t_{U'}) = Y_{U'}(t_{U'}) + \sum_{\tilde{U} \in \mathbb{P}(U \setminus U')} (-1)^{|\tilde{U}|} \lim_{t_{\tilde{U}} \to \infty} Y_{\tilde{U} \cup U'}(t_{\tilde{U} \cup U'})$$

$$= \sum_{\tilde{U} \in \mathbb{P}(\bar{U} \setminus U')} (-1)^{|\tilde{U}|} \lim_{t_{\tilde{U}} \to \infty} Y_{\tilde{U} \cup U'}(t_{\tilde{U} \cup U'})$$

$$= \sum_{\substack{\bar{U} \in \mathbb{P}(U) \\ \bar{U} \supseteq U'}} (-1)^{|\bar{U} \setminus U'|} \lim_{t_{\bar{U} \setminus U'} \to \infty} Y_{\bar{U}}(t_{\bar{U}})$$

If the GERT network in question is only weakly admissible (that is, the assumptions A5 and A6 need not be satisfied), the activation functions Y_U can again be defined by (2.1.2). Relations (2.1.5) and (2.1.6) and definition (2.1.7), however, are no longer valid because, for a weakly admissible GERT network, nodes outside cycles may be activated several times during a single project execution.

2.2 Evaluation of Admissible GERT Networks

As we will see in section 2.3, some quantities important to time planning of projects can be determined from $Y(S)$ or $\mathcal{Y}(S)$, respectively, where S is again the sink set of the admissible GERT network under consideration. We therefore define

(2.2.1)   **Definition.**
By the **complete evaluation** of an admissible GERT network N (in the sense of time planning) we mean the determination of the family of activation functions $Y(S)$ for the sink set S of N.

(2.2.2) Note that it also suffices to compute the family of activation distributions $\mathcal{Y}(S)$ in order to evaluate an admissible GERT network completely. Since we only consider activation distributions of sets $S' \in \mathbb{P}(U)$ with respect to the fixed set U=S from now on, we write $\mathcal{Y}_{S'}$ instead of $\mathcal{Y}^S_{S'}$ in what follows.

Next we take a closer look at the complete evaluation of an admissible GERT network N with several sources. Let R be again the source set of N and let

$$S'(i) := S' \cap \mathcal{R}(i) \quad \text{for } i \in R' \in \hat{\mathbb{P}}(R), \ S' \in \mathbb{P}(S)$$

Since R' is a feasible subset of R (that is, $\mathcal{R}(i) \cap \mathcal{R}(k) = \emptyset$ for $i,k \in R', i \neq k$), we have $S'(i) \cap S'(k) = \emptyset$ for $i,k \in R', i \neq k$. Let

(2.2.3) $$\tilde{\mathbb{P}}(R) := \{R' \in \hat{\mathbb{P}}(R) \mid q_{R'} > 0\}$$

Then for $R' \in \tilde{\mathbb{P}}(R)$, $S' \in \mathbb{P}(\underset{i \in R'}{\cup} S(i))$

$$(2.2.4)\ Y_{R',S'}(\tau_{R'},t_{S'}) := \begin{cases} P[\underset{l \in R'}{\cap}\ \underset{j \in S'(1)}{\cap}\ (T_j \leq t_j) \mid (T_i = \tau_i)_{i \in R'} \cap (\mathcal{A}_k)_{k \in R \setminus R'}], \\ \qquad \text{if } t_j \geq \tau_i \text{ for all } j \in S'(i) \text{ and } i \in R' \\ 0, \text{ otherwise} \end{cases}$$

is the conditional probability that each sink $j \in S'$ is activated by time t_j given that each source $i \in R'$ has been activated at time τ_i and none of the remaining sources $k \in R \setminus R'$ has been activated. The mapping $Y_{R',S'} : \mathbb{R}^{|R'|+|S'|} \to [0,1]$ defined by (2.2.4) is called the **activation function of sink subset** S' **given exactly the source subset** R'. Similarly, we can define the activation distributions $y_{R',S'}$ (cf. Neumann (1983)).

Now we show that because of assumptions A1 and A2a we need only compute the activation functions $Y_{i,S'}$ for $i \in R'$, $S' \in \mathbb{P}(S(i))$ in order to determine the functions $Y_{R',S'}$ $(R' \in \tilde{\mathbb{P}}(R), S' \in \mathbb{P}(S))$. First we put $Y_{R',S'} := 0$ if there is a sink in S' which is not reachable from any source from R' :

$$Y_{R',S'} := 0 \text{ if there is a } j \in S' \text{ such that } R' \cap \mathbb{Z}(j) = \emptyset$$

If $R' \cap \mathbb{Z}(j) \neq \emptyset$ for all $j \in S'$, we obtain by assumptions A1 and A2a

$$Y_{R',S'}(\tau_{R'},t_{S'}) = \underset{i \in R'}{\Pi}\ Y_{i,S'(i)}(\tau_i,t_{S'(i)})$$

(compare remark (1.3.6a)) where

$$(2.2.5) \qquad Y_{i,S'(i)}(\tau_i,t_{S'(i)}) := 1 \quad \text{for } S'(i) = \emptyset$$

is the probability that no or at least one sink is activated given that exactly the source i has been activated (at time τ_i). In summary, we have

$$
(2.2.6) \qquad Y_{R',S'}(\tau_{R'},t_{S'}) = \begin{cases} 0, & \text{if } R' \cap \overline{Z}(j)=\emptyset \quad \text{for some } j \in S' \\[2mm] \underset{i \in R'}{\Pi} \ Y_{i,S'(i)}(\tau_i,t_{S'(i)}), & \text{otherwise} \end{cases}
$$

Because of the stationariness property from remark (1.3.6d) it holds that

$$
Y_{i,S'(i)}(\tau_i,t_{S'(i)}) = Y_{i,S'(i)}(0,t_{S'(i)}-\tau_i)
$$

where $t_{S'(i)}-\tau_i$ means the vector with the components $t_j-\tau_i$ ($j \in S'(i)$). For simplicity we write $Y_{i,\overline{S}}(t_{\overline{S}})$ instead of $Y_{i,\overline{S}}(0,t_{\overline{S}})$ in what follows, that is

$$
(2.2.7) \qquad Y_{i,\overline{S}}(t_{\overline{S}}) := \begin{cases} P[(T_j \le t_j)_{j \in \overline{S}} | A_i \cap (\overline{A}_k)_{k \in R \setminus \{i\}}] & \text{for } t_j \in \mathbb{R}_+, \ j \in \overline{S} \in \overline{P}(S(i)) \\[2mm] 0, & \text{otherwise} \end{cases}
$$

Then (2.2.6) becomes

$$
(2.2.8) \qquad Y_{R',S'}(\tau_{R'},t_{S'}) = \begin{cases} 0, & \text{if } R' \cap \overline{Z}(j)=\emptyset \quad \text{for some } j \in S' \\[2mm] \underset{i \in R'}{\Pi} \ Y_{i,S'(i)}(t_{S'(i)}-\tau_i), & \text{otherwise} \end{cases}
$$

$$
\text{for } R' \in \tilde{P}(R), \ S' \in P(S)
$$

Recall the theorem of total probability, which says the following: Let $B_1,\ldots,B_r$ be disjoint random events with $P(B_\rho)>0$ $(\rho=1,\ldots,r)$ and $P(\underset{\rho=1}{\overset{r}{\cup}} B_\rho)=1$. Then we have for an arbitrary random event A

$$
P(A) = \underset{\rho=1}{\overset{r}{\Sigma}} P(A|B_\rho)P(B_\rho)
$$

The following generalization of the theorem of total probability holds to be true: Let X and Z be random variables with ranges $\mathbb{R}^m$ and $\mathbb{R}^n$, respectively, where $\mathbb{R}:=\mathbb{R}\cup\{\infty\}$, and let Φ be the distribution function of Z. Furthermore, let $C \subseteq \mathbb{R}^m$ and $D \subseteq \mathbb{R}^n$ be measurable sets. Then

(2.2.9) $P[(X \in C) \cap (Z \in D)] = \int_D P(X \in C | Z = z) \Phi(dz)$

Using (2.1.6), (2.2.4) and (2.2.9) we obtain

(2.2.10) $Y_{S'}(t_{S'}) = \sum\limits_{R' \in \tilde{\mathbb{P}}(R)} \int\limits_{\mathbb{R}_+^{|R'|}} Y_{R',S'}(\tau_{R'}, t_{S'}) \varkappa_{R'}(d\tau_{R'})$ for $S' \in \mathbb{P}(S)$

(2.2.10) and (2.2.8) show that it suffices to compute the activation functions $Y_{i,S}(\cdot)$ for $i \in R$, $S \in \mathbb{P}(S \cap \mathcal{R}(i))$ in order to evaluate the underlying admissible GERT network with source set R completely. We also see that the family of activation functions $Y(S)$ of an admissible GERT network with several sources and of the corresponding one–source network (cf. (1.3.4)) coincide.

Next we will see that for completely evaluating an admissible GERT network which satisfies the so–called one–sink condition it suffices to compute the activation functions Y_{ij} defined as

$$Y_{ij}(t) := \begin{cases} P[T_j \leq t | A_i \cap (\overline{A}_k)_{k \in R \setminus \{i\}}] & \text{for } t \geq 0 \\ 0 & \text{for } t < 0 \end{cases} \qquad (i \in R, \ j \in S \cap \mathcal{R}(i))$$

provided that $q_i > 0$ (compare (2.2.7)). Let $i \in R$ and $j \in S \cap \mathcal{R}(i)$. If i is activated during a project execution, then owing to assumption A1 no additional source k from which sink j is reachable is activated during that project execution. Hence,

$$P[T_j \leq t | A_i \cap (\overline{A}_k)_{k \in R \setminus \{i\}}] = P[T_j \leq t | A_i] = P(T_j \leq t | T_i = 0)$$

and

(2.2.11) $Y_{ij}(t) = \begin{cases} P(T_j \leq t | T_i = 0) & \text{for } t \geq 0 \\ 0 & \text{for } t < 0 \end{cases} \qquad (i \in R, \ j \in S \cap \mathcal{R}(i))$

provided that $q_i > 0$. Y_{ij} is called the **activation function of sink j given source** i.

(2.2.12) **Definition.**

A GERT network N is said to satisfy the **one–sink condition** if, for every source i of N, during each execution of the GERT subnetwork N(i) induced by $\mathcal{R}(i)$ (cf. (1.4.5)) exactly one sink of N(i) is activated with probability 1.

Note that an execution of N(i) starts when the source i of N(i) is activated during an execution of N. For a GERT network N which satisfies the one–sink condition, the number of sinks activated during any execution of N is, with probability 1, equal to the number of sources activated during that project execution. As we will see in section 3.2, each STEOR network satisfies the one–sink condition.

(2.2.13) Note that for a GERT network with several sources which satisfies the one–sink condition, the corresponding one–source network does not generally fulfill the one–sink condition unless all feasible subsets $R' \in \hat{\mathbb{P}}(R)$ are singletons.

For an admissible GERT network with only one source that satisfies the one–sink condition we have

$$(2.2.14) \quad \begin{cases} Y_j = \mathcal{Y}_j & \text{for all } j \in S \\ Y_{S'} = \mathcal{Y}_{S'} = 0 & \text{for all } S' \in \mathbb{P}(S) \text{ with } |S'| > 1 \end{cases}$$

Now we consider an admissible GERT network with several sources that satisfies the one–sink condition. The latter condition implies

$$Y_{R',S'} = 0 \quad \text{for } |R'| < |S'|$$

in particular, $Y_{i,S'(i)} = 0$ for $|S'(i)| > 1$, where again $S'(i) := S' \cap \mathcal{R}(i)$. Observing (2.2.5) relation (2.2.8) takes the form

$$(2.2.15) \qquad Y_{R',S'}(\tau_{R'},t_{S'}) = \begin{cases} 0, \text{ if } R' \cap \mathbf{Z}(j)=\emptyset \text{ for some } j \in S' \\ \qquad \text{or } |S'(i)|>1 \text{ for some } i \in R' \\ \prod_{\substack{i \in R' \\ |S'(i)|=1}} Y_{i,S'(i)}(t_{S'(i)} - \tau_i), \text{ otherwise} \end{cases}$$

$$\text{for } R' \in \tilde{R}(R), \ S' \in \mathbf{P}(S)$$

(2.2.16) Remarks.

(a) From (2.2.10) and (2.2.15) we see that for completely evaluating an admissible GERT network N which satisfies the one–sink condition, we need only compute the activation functions Y_{ij} $(i \in R, j \in S \cap \mathbf{Z}(i))$. If N contains only one source, the computation of the activation functions Y_j $(j \in S)$ is sufficient for the complete evaluation of N.

(b) The complete evaluation of an admissible GERT network is in general very complicated and time consuming (as we will see in section 2.4) unless the one–sink condition is satisfied. Even if this condition is not fulfilled, we will often restrict ourselves to the simpler computation of the activation functions Y_{ij} $(i \in R, j \in S \cap \mathbf{Z}(i))$. In the latter case, we will speak of the **evaluation** of the admissible GERT network in question (without "complete").

(c) All the preceding definitions and relations concerning the activation functions $Y_{R',S'}$ remain valid if the GERT network in question is only weakly admissible and we replace the conditional probabilities by the corresponding conditional expected values.

In evaluating an admissible GERT network, sometimes it turns out to be expedient to reduce the network or parts of it to "smaller" (sub)networks whose evaluation is less time consuming. Then it must be guaranteed that the families of activation functions Y(S) for the sink sets of both the original and the modified GERT network coincide. This leads to the following

(2.2.17)   Definition.

Two weakly admissible GERT networks N_1 and N_2 with node sets V_1 and V_2, respectively, are said to be **equivalent** if the following conditions are met:

(a) N_1 and N_2 have the same sink set S, and the families of activation functions $Y(S)$ of N_1 and N_2 coincide.

(b) Let $\mathcal{R}_\nu(i)$ be the set of those nodes in N_ν which are reachable from $i \in V_\nu$ $(\nu=1,2)$. Then $\mathcal{R}_1(i) \cap V_2 = \mathcal{R}_2(i) \cap V_1$ for all $i \in V_1 \cap V_2$.

(2.2.18)   Remarks.

(a) In definition (2.2.17) the sinks in N_1 need not be of the same type as the corresponding sinks in N_2.

(b) Condition (2.2.17b) says that if, for two nodes i and j which belong to both N_1 and N_2, j is reachable from i in N_1, then j is also reachable from i in N_2 and vice versa. Thus, condition (2.2.17b) stipulates the "conservation of reachability". For example, this condition is not satisfied if N_1 and N_2 differ in the dummy activity $\langle i,j \rangle$ shown in Fig. 2.2.1.

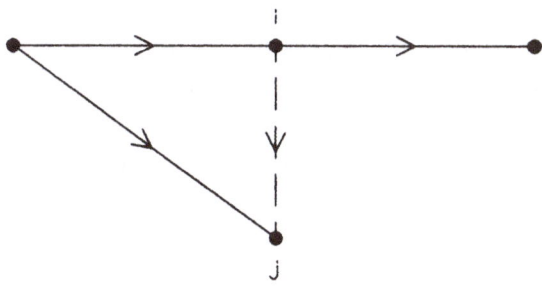

Figure 2.2.1

(c) A GERT network N_2 which is equivalent to an admissible GERT network N_1 is not necessarily admissible as it is shown in Fig. 2.2.2, where each activity $\langle i,j \rangle$ is marked with its (deterministic) duration D_{ij} and where node k in N_2 violates assumption A5.

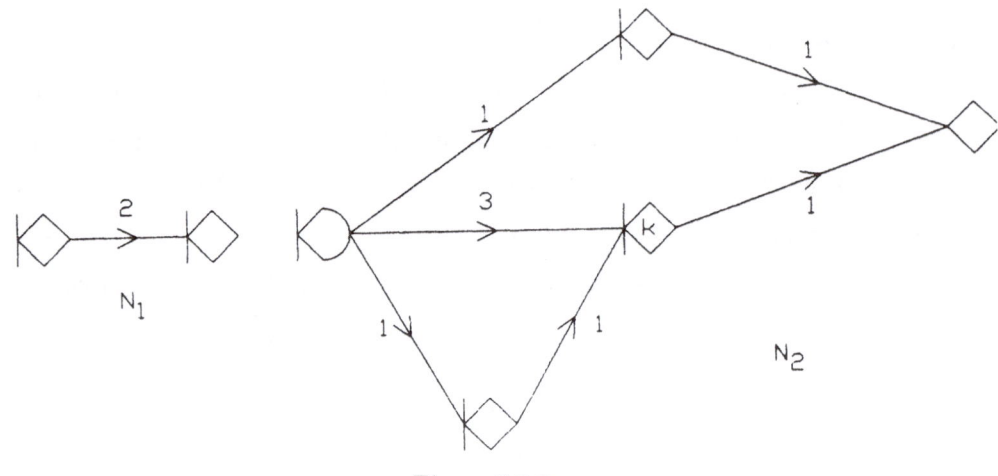

Figure 2.2.2

(d) The addition of dummy activities necessary to satisfy conventions (1.4.1) and
(1.4.2) for a complete subnetwork N' of a weakly admissible GERT network N as it
is shown in Figs. 1.4.2 and 1.4.3 leads to a GERT network equivalent to the
original network N.

(e) A weakly admissible GERT network with several sources is equivalent to its
corresponding one–source network.

2.3 Computation of Some Quantities Important to Time Planning

We show that some important quantities related to the sink set S of an admissible
GERT network can be determined from $Y(S)$ and $\mathcal{Y}(S)$.

The probability $q_{S'}$ that (at least) all sinks $j \in S' \subseteq S$ are activated (during a project
execution) is

$$q_{S'} := P[(\mathcal{A}_j)_{j \in S'}] = \lim_{t_{S'} \to \infty} Y_{S'}(t_{S'}) \quad \text{for } S' \neq \emptyset$$

The probability $q_{S'}$ that exactly all sinks $j \in S' \subseteq S$ are activated is

$$q_{S'} := P[(\mathcal{A}_j)_{j \in S'} \cap (\bar{\mathcal{A}}_i)_{i \in S \setminus S'}] = \lim_{t_{S'} \to \infty} \mathcal{Y}_{S'}(t_{S'}) \quad \text{for } S' \neq \emptyset$$

Trivially,

$$q_{S'} = 1 \quad \text{for } S' = \emptyset$$

and

$$q_S = q_S$$

In analogy to (1.2.13) and (1.2.15) we have

(2.3.1) $$q_{S'} = \sum_{\substack{\bar{S} \in \mathbb{P}(S) \\ \bar{S} \supseteq S'}} q_{\bar{S}}$$

$$q_{S'} = \sum_{\substack{\bar{S} \in \mathbb{P}(S) \\ \bar{S} \supseteq S'}} (-1)^{|\bar{S} \setminus S'|} q_{\bar{S}} \qquad \text{for } S' \in \mathbb{P}(S)$$

Moreover,

$$q := \sum_{S' \in \mathbb{P}(S)} q_{S'}$$

is the probability that at least one sink is activated and

$$q_{S'} = 1 - q \quad \text{for } S' = \emptyset$$

is the probability that no sink is activated. Fig. 2.3.1 shows an admissible GERT network (where the arc weights are omitted) with q=0.

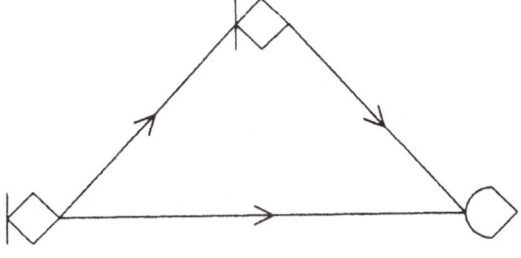

Figure 2.3.1

If the GERT network has only one source and the one–sink condition is satisfied, it holds that

$$q_j = q_j \qquad \text{for all } j \in S$$
$$q_{S'} = q_{S'} = 0 \qquad \text{for } |S'| > 1$$

The conditional distribution function $G_{S'}$ of the project duration given that (at least) all sinks $j \in S' \neq \emptyset$ are activated is

$$G_{S'}(t_{S'}) := P[(T_j \leq t_j)_{j \in S'} \mid (A_j)_{j \in S'}] = \frac{Y_{S'}(t_{S'})}{q_{S'}}$$

provided that $q_{S'} > 0$. The conditional distribution function $G_{S'}^{skip}$ of the time up to the activation of the last (with respect to time) of the sinks $j \in S' \neq \emptyset$ (that is, the conditional distribution function of the skipping project duration) given that (at least) the sinks $j \in S'$ are activated becomes

$$G_{S'}^{skip}(t) := P[(T_j \leq t)_{j \in S'} \mid (A_j)_{j \in S'}] = \frac{Y_{S'}[(t_j = t)_{j \in S'}]}{q_{S'}}$$

provided that $q_{S'} > 0$.

The distribution function $G_{S'}$ does not really describe the (nonskipping) project duration, that is, the earliest point in time after which no more activities are being carried out. The latter quantity cannot be expressed in terms of $Y(S)$ alone. Nevertheless, the definition of $G_{S'}$ makes some sense because, in contrast to the skipping project duration, the (nonskipping) project duration does not only depend on the time of activation of the "latest" sink from S'.

The conditional distribution function $\mathcal{G}_{S'}$ of the project duration given that exactly all sinks $j \in S' \neq \emptyset$ are activated is

$$\mathcal{G}_{S'}(t_{S'}) := P[(T_j \leq t_j)_{j \in S'} \mid (A_j)_{j \in S'} \cap (\overline{A}_i)_{i \in S \setminus S'}]$$

$$= \frac{\mathcal{Y}_{S'}(t_{S'})}{q_{S'}}$$

provided that $q_{S'} > 0$, where we have used the relation $P(A \mid B \cap C) = P(A \cap C \mid B \cap C)$. The conditional distribution function $\mathcal{G}_{S'}^{skip}$ of the skipping project duration given that exactly all sinks $j \in S' \neq \emptyset$ are activated becomes

$$\mathcal{G}_{S'}^{skip}(t) := P[(T_j \leq t)_{j \in S'} \mid (A_j)_{j \in S'} \cap (\overline{A}_i)_{i \in S \setminus S'}]$$

$$= \frac{\mathcal{Y}_{S'}[(t_j = t)_{j \in S'}]}{q_{S'}}$$

provided that $q_{S'} > 0$.

Obviously,

$$\left. \begin{array}{l} G_j = G_j^{skip} \\[2mm] \mathcal{G}_j = \mathcal{G}_j^{skip} \end{array} \right\} \text{ for all } j \in S$$

If the GERT network has only one source and the one–sink condition is satisfied, then

$$G_j = \mathcal{G}_j = G_j^{skip} = \mathcal{G}_j^{skip} \quad \text{for all } j \in S$$

Moreover, for $|S'| > 1$, we have $q_{S'} = q_{S'} = 0$, and $\mathcal{G}_{S'}$, $G_{S'}^{skip}$, $\mathcal{G}_{S'}$ and $\mathcal{G}_{S'}^{skip}$ are thus not defined.

At last we determine the (unconditional) distribution function G^{skip} of the skipping project duration D^{skip} provided that $q = 1$. Let $S_1, S_2, \ldots$ be the nonempty subsets of S. Then $D^{skip} \leq t$ means that exactly all sinks $j \in S_1$ are activated and $T_j \leq t$ for all $j \in S_1$ or that exactly all sinks $j \in S_2$ are activated and $T_j \leq t$ for all $j \in S_2$ or etc. Since the random events in question are disjoint, we obtain

$$(2.3.2) \qquad G^{skip}(t) := P(D^{skip} \leq t) = \sum_{S' \in \mathbb{P}(S)} P[(T_j \leq t)_{j \in S'} \cap (\overline{A}_i)_{i \in S \setminus S'}]$$

$$= \sum_{S' \in \mathbb{P}(S)} Y_{S'}[(t_j = t)_{j \in S'}]$$

If the GERT network has only one source and satisfies the one–sink condition, we have by (2.3.2) and (2.2.14)

$$(2.3.3) \qquad G^{skip} = \sum_{j \in S} Y_j = \sum_{j \in S} q_j G_j$$

As an example, we consider the admissible GERT network in Fig. 2.3.2 where the execution probabilities of the activities emanating from stochastic nodes are written onto the respective arcs (the distribution functions of the activity durations are omitted).

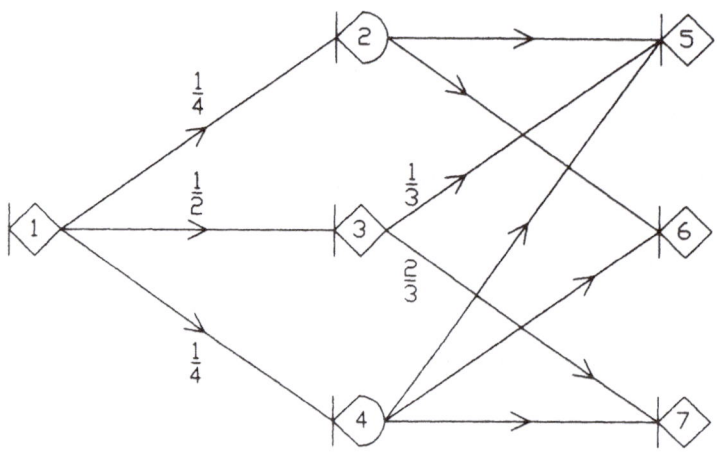

Figure 2.3.2

The sink set is $S=\{5,6,7\}$. The probabilities $q_{S'}$ and $q_{S'}$ for all $S' \in P(S)$ are listed in Table 2.3.1, where the $q_{S'}$ are found by using (2.3.1). The probability that at least one sink is activated is

$$q = \sum_{S' \in P(S)} q_{S'} = 1$$

S'	∅	{5}	{6}	{7}	{5,6}	{5,7}	{6,7}	{5,6,7}
$q_{S'}$	0	$\frac{1}{6}$	0	$\frac{1}{3}$	$\frac{1}{4}$	0	0	$\frac{1}{4}$
$q_{S'}$	1	$\frac{2}{3}$	$\frac{1}{2}$	$\frac{7}{12}$	$\frac{1}{2}$	$\frac{1}{4}$	$\frac{1}{4}$	$\frac{1}{4}$

Table 2.3.1

2.4 Evaluation Methods for Admissible GERT Networks

To evaluate an admissible GERT network N completely it is in general expedient to reduce N first to an acyclic admissible GERT network which is equivalent to N. This "cycle reduction" will be discussed in section 4.2.

Steinhardt has developed a procedure for computing the family of activation distributions $\mathcal{Y}(S)$ for the sink set S of an admissible acyclic GERT network N, which will briefly be sketched in what follows (for details we refer to Neumann and Steinhardt (1979a), section 4.2). Steinhardt's method constructs a sequence $N^1,\dots,N^m$ of subgraphs of N where N^1 consists of the sources of N, N^m coincides with N, and $N^2,\dots,N^{m-1}$ are subnetworks of N which represent admissible GERT networks when treated as separate entities. A subnetwork N^μ ($2 \leq \mu \leq m$) results from the preceding subnetwork $N^{\mu-1}$ by adding to $N^{\mu-1}$ at least one node $j \in V \backslash V^{\mu-1}$ with $P(j) \subseteq V^{\mu-1}$ and all incoming arcs $<i,j>$, $i \in P(j)$, where V and $V^{\mu-1}$ are the node sets of N and $N^{\mu-1}$, respectively. Then Steinhardt's method successively determines $\mathcal{Y}(S^2),\dots,\mathcal{Y}(S^m)$ where S^μ is the sink set of N^μ ($\mu=2,\dots,m$). $\mathcal{Y}(S^\mu)$ can be computed from $\mathcal{Y}(S^{\mu-1})$ as follows, where we write $\mathcal{Y}_U^\mu$ instead of $\mathcal{Y}_U^{S^\mu}$:

(2.4.1) $$\mathcal{Y}_U^\mu(t_U) = \sum_{U' \in \mathbb{P}(S^{\mu-1})} \int_{\mathbb{R}_+^{|U'|}} P[(T_j \leq t_j)_{j \in U} \cap (\mathcal{A}_i)_{i \in S^\mu \setminus U}$$

$$| (T_j \leq \tau_j)_{j \in U'} \cap (\mathcal{A}_i)_{i \in S^{\mu-1} \setminus U'}] \mathcal{Y}_{U'}^{\mu-1}(d\tau_{U'}) \qquad \text{for } U \in \mathbb{P}(S^\mu)$$

The choice of the subnetwork N^μ is such that the computation of the conditional probabilities in (2.4.1) is not too complicated. The evaluation of (2.4.1), however, requires the evaluation of multiple integrals. We note that, in the course of Steinhardt's procedure, it is also tested whether N satisfies the "hard" assumption A6.

In step μ of Steinhardt's method $(2 \leq \mu \leq m)$, all activation distributions $\mathcal{Y}_{U_\mu}^\mu$ with $U_\mu \in \mathbb{P}(S^\mu)$ are computed from the functions $\mathcal{Y}_{U_{\mu-1}}^{\mu-1}$ $(U_{\mu-1} \in \mathbb{P}(S^{\mu-1}))$. For finding the functions $\mathcal{Y}_{U_\mu}^\mu$, however, not necessarily all functions $\mathcal{Y}_U^{\mu-1}$ are needed. Nicolai (1981) has proposed a method for computing the family of activation functions $Y(S)$, which determines a sequence $M_1, \ldots, M_k$ of subsets of $\mathbb{P}(V)$ with $\mathbb{P}(S) \subseteq \bigcup_{\kappa=1}^k M_\kappa$ and, in general, $|\bigcup_{\kappa=1}^k M_\kappa| < |\bigcup_{\mu=1}^m \mathbb{P}(S^\mu)|$. The activation functions Y_{U_κ} for $U_\kappa \in M_\kappa$ can be computed from the functions $Y_{U_{\kappa-1}}$ $(U_{\kappa-1} \in M_{\kappa-1})$ by formulas similar to (2.4.1), where all $Y_{U_{\kappa-1}}$ are needed $(\kappa=2, \ldots, k)$.

In Neumann (1985a), an algorithm for computing the family of activation functions $Y(S)$ of admissible EOR networks (that is, GERT networks all of whose nodes have EOR entrance and deterministic or stochastic exit) is given, which is much less complicated than Nicolai's method. All three procedures mentioned in this section, however, require the evaluation of multiple integrals and are thus very time consuming and should be used only for smaller GERT networks. In practice, simulation methods are widely used for evaluating general GERT networks (see Neumann and Steinhardt (1979a), chapter 7, Pritsker (1977), Pritsker and Sigal (1983), and Whitehouse (1973)).

In chapters 3 and 4, we will discuss methods for computing the activation functions Y_{ij} for STEOR networks, EOR networks, and so–called BES networks (which are equivalent to STEOR networks or, respectively, EOR networks). Unless the one–sink condition is satisfied (which is the case for STEOR networks and for BES networks equivalent to STEOR networks), those methods do not provide the complete evaluation of the respective GERT networks. A combination of those methods and the simulation technique turns out to be expedient for evaluating general GERT networks (not completely!) as we will see in section 4.6.

Chapter 3   STEOR Networks and EOR Networks

STEOR networks are GERT networks with only STEOR nodes. STEOR networks represent that class of GERT networks whose evaluation is the easiest due to the fact that Markov renewal processes can be associated with those networks. In particular, STEOR networks satisfy the one–sink condition. Exploiting some results from the theory of Markov renewal processes, an efficient method for computing the activation functions Y_{ij} can be found.

EOR networks are GERT networks all of whose nodes have an EOR entrance but which may possess a stochastic or a deterministic exit. As we will see, the activation functions Y_{ij} for EOR networks can be computed in the same way as for STEOR networks. Since an EOR network that does not represent a STEOR network does not satisfy the one–sink condition, the activation functions Y_{ij} do not yield the complete evaluation of the EOR network.

Before we turn to STEOR networks, we will summarize the most important concepts and results from the theory of Markov chains and Markov renewal processes which are needed for what follows. For details we refer to Cinlar (1975).

3.1 Markov Chains and Markov Renewal Processes

A family $\{X(s) \mid s \in I\}$ of random variables is called a **stochastic process**. In what follows, we only consider the case $I = \mathbb{N}_0$, and we then write $(X_\nu)_{\nu \in \mathbb{N}_0}$ instead of $\{X(s) \mid s \in \mathbb{N}_0\}$.

Each of the random variables X_ν $(\nu \in \mathbb{N}_0)$ is supposed to be a mapping $X_\nu : \Omega \to \mathcal{E}$ of one and the same sample space Ω into one and the same set $\mathcal{E}$. The set $\mathcal{E}$ is called the **state space** of the stochastic process $(X_\nu)_{\nu \in \mathbb{N}_0}$, and the elements of $\mathcal{E}$ are the **states**.

A stochastic process $(X_\nu)_{\nu \in \mathbb{N}_0}$ is said to be a **Markov chain** if for all $\nu \in \mathbb{N}_0$ and all $x_0, x_1, \ldots, x_{\nu+1} \in \mathcal{E}$ it holds that

(3.1.1) $P(X_{\nu+1}=x_{\nu+1}|X_\nu=x_\nu,\ldots,X_0=x_0) = P(X_{\nu+1}=x_{\nu+1}|X_\nu=x_\nu)$

Instead   of $P(X_{\nu+1}=x_{\nu+1}|X_\nu=x_\nu,\ldots,X_0=x_0)$,   we   use   the   shorter   form
$P(X_{\nu+1}=x_{\nu+1}|X_\nu,\ldots,X_0)$ in what follows. Condition (3.1.1) is referred to as a **Markov property**, which says that the "next" state $X_{\nu+1}$ of the process is conditionally independent of the "past" states $X_0,\ldots,X_{\nu-1}$ provided that the "present" state X_ν is known (compare the Markov property implied by assumption A2d).

The quantities

$$p_{ij}(\nu) := P(X_{\nu+1}=j|X_\nu=i)   \quad (i,j\in\mathcal{E}; \nu\in\mathbb{N}_0)$$

are called the **transition probabilities** of the Markov chain $(X_\nu)_{\nu\in\mathbb{N}_0}$. If these transition probabilities are independent of ν for all $i,j\in\mathcal{E}$ (we then write p_{ij} instead of $p_{ij}(\nu)$), $(X_\nu)_{\nu\in\mathbb{N}_0}$ is called a **homogeneous Markov chain**. In what follows, we will only consider homogeneous Markov chains with state space $\mathcal{E}\subseteq\mathbb{N}_0$. A homogeneous Markov chain $(X_\nu)_{\nu\in\mathbb{N}_0}$ is uniquely specified once the transition probabilities p_{ij} $(i,j\in\mathcal{E})$ and the **initial distribution** given by the probabilities $p_i:=P(X_0=i)$, $i\in\mathcal{E}$, are known.

The transition probabilities satisfy the relation

$$\sum_{j\in\mathcal{E}} p_{ij} = 1 \quad (i\in\mathcal{E})$$

For the so-called μ-**step transition probabilities**

$$p_{ij}^{(\mu)} := P(X_\mu=j|X_0=i) \quad (\mu\in\mathbb{N})$$

we have the corresponding relation

(3.1.2) $\sum\limits_{j\in\mathcal{E}} p_{ij}^{(\mu)} = 1$ $(i\in\mathcal{E}, \mu\in\mathbb{N})$

and the **Chapman–Kolmogorov equation**

(3.1.3) $p_{ij}^{(\mu+\nu)} = \sum\limits_{k\in\mathcal{E}} p_{ik}^{(\mu)} p_{kj}^{(\nu)}$ $(i, j\in\mathcal{E}; \mu, \nu\in\mathbb{N})$

Now let

$$f_{ij}^{(\mu)} := P(X_\mu = j, X_\nu \neq j \ (\nu = 1, \dots, \mu-1) \mid X_0 = i)   (i, j\in\mathcal{E}; \mu\in\mathbb{N})$$

be the probability that starting at state i the state j is entered for the first time after μ "transitions", and let

$$f_{ij}^* := \sum\limits_{\mu=1}^{\infty} f_{ij}^{(\mu)}   (i, j\in\mathcal{E})$$

be the probability that state j can be reached at all from state i. Then state i is said to be

recurrent	if $f_{ii}^* = 1$
transient	if $f_{ii}^* < 1$

A special recurrent state is the so–called **absorbing state** which is determined by the condition $p_{ii} = 1$.

We now present two theorems concerning the types of states just introduced.

(3.1.4)   **Theorem.**
Let i be a recurrent state and j be a transient state. Then $p_{ij}^{(\mu)} = 0$ for all $\mu\in\mathbb{N}$.

Theorem (3.1.4) says that, with positive probability, one can only move from a recurrent state to another recurrent state. Let M_j be the number of times state j occurs during the evolution of the stochastic process in question. Then we have

(3.1.5)   Theorem.

For a transient state j it holds that

(a) $\lim\limits_{\mu \to \infty} p_{ij}^{(\mu)} = 0$

$\left. \right\}$ for every $i \in \mathcal{E}$

(b) $\mathbb{E}(M_j \mid X_0 = i) < \infty$ and thus $P(M_j < \infty \mid X_0 = i) = 1$

(3.1.5a) says that the stochastic process will eventually terminate in a recurrent state. In (3.1.5b) $\mathbb{E}(M_j \mid X_0 = i)$ is the conditional expected value of random variable M_j given that the stochastic process starts at state i [4].

Next, we introduce the concept of a Markov renewal process. A stochastic process $(X_\nu, \theta_\nu)_{\nu \in \mathbb{N}_0}$ with **state space** $\mathcal{E} \times \mathbb{R}_+$ [5] and $\mathcal{E} \subseteq \mathbb{N}_0$ is called a **Markov renewal process** if

(3.1.6)
$\left[\begin{array}{l} \text{(a) } 0 = \theta_0 \leq \theta_1 \leq \theta_2 \leq \ldots \\[1mm] \text{(b) } P(X_{\nu+1} = j, \theta_{\nu+1} - \theta_\nu \leq t \mid X_\nu, \ldots, X_0; \theta_\nu, \ldots, \theta_0) \\[1mm] \qquad = P(X_{\nu+1} = j, \theta_{\nu+1} - \theta_\nu \leq t \mid X_\nu) \quad \text{for all } \nu \in \mathbb{N}_0, j \in \mathcal{E}, \text{ and } t \geq 0 \end{array} \right.$

A Markov renewal process describes the evolution in time of a system taking the states $X_0, X_1, X_2, \ldots$ consecutively at the time points $\theta_0 = 0, \theta_1, \theta_2, \ldots$ where (3.1.6b) holds. Instead of "occurrence of a state" we also speak of the "renewal of a state" in what follows. If $(X_\nu, \theta_\nu)_{\nu \in \mathbb{N}_0}$ is a Markov renewal process, then $(X_\nu)_{\nu \in \mathbb{N}_0}$ is called the **imbedded Markov chain.**

---

[4]   If j=i, the start at state i is considered to be the first occurrence of state i.

[5]   In dealing with Markov renewal processes, sometimes the pairs $(i,t) \in \mathcal{E} \times \mathbb{R}_+$ as well as the elements $i \in \mathcal{E}$ are referred to as "states".

A Markov renewal process is said to be **homogeneous** if the quantities

(3.1.7) $P(X_{\nu+1}=j,\theta_{\nu+1}-\theta_{\nu}\le t\,|\,X_{\nu}=i) =: \mathbb{Q}_{ij}(t)$ $(i,j\epsilon\mathcal{E};t\ge 0)$

do not depend on ν. The functions $\mathbb{Q}_{ij}$ given by (3.1.7) and $\mathbb{Q}_{ij}(t):=0$ for $t<0$ are called the **transition functions** of the homogeneous Markov renewal process. In what follows, we only consider homogeneous Markov renewal processes. A homogeneous Markov renewal process $(X_{\nu},\theta_{\nu})_{\nu\epsilon\mathbb{N}_0}$ is uniquely specified once the transition functions $\mathbb{Q}_{ij}$ $(i,j\epsilon\mathcal{E})$ and the **initial distribution** given by the probabilities $p_i:=P(X_0=i)$, $i\epsilon\mathcal{E}$, are known.

Let $M_j(t)$ be the number of renewals of state j in the time interval $[0,t]$. The functions R_{ij} defined by

(3.1.8) $R_{ij}(t) := \begin{cases} \mathbb{E}[M_j(t)\,|\,X_0=i] & \text{for } t\ge 0 \\ 0 & \text{for } t<0 \end{cases}$ $(i,j\epsilon\mathcal{E})$

are called the **renewal functions** of the respective Markov renewal process. According to definition, the functions R_{ij} are increasing. By theorem (3.1.5b), the functions R_{ij} are bounded if j is a transient state.

Finally, we introduce the μ–**step transition functions** $\mathbb{Q}_{ij}^{(\mu)}$ by

(3.1.9) $\mathbb{Q}_{ij}^{(\mu)}(t) := \begin{cases} P(X_{\mu}=j,\theta_{\mu}\le t\,|\,X_0=i) & \text{for } t\ge 0 \\ 0 & \text{for } t<0 \end{cases}$ $(i,j\epsilon\mathcal{E};\mu\epsilon\mathbb{N}_0)$

For $\mu=0$ we have

(3.1.10) $\mathbb{Q}_{ij}^{(0)}(t) - \delta_{ij}(t) := \begin{cases} 1, & \text{for } i=j \text{ and } t\ge 0 \\ 0, & \text{otherwise} \end{cases}$

Moreover, $Q_{ij}^{(1)} = Q_{ij}$.

The μ-step transition functions $Q_{ij}^{(\mu)}$, like distribution functions, are increasing and continuous from the right and can be computed successively as follows:

$$(3.1.11) \qquad Q_{ij}^{(\mu)}(t) = \sum_{k \in \mathcal{E}} \int_{[0,t]} Q_{ik}^{(\mu-1)}(t-s) \, Q_{kj}(ds) \qquad (i,j \in \mathcal{E}; \mu \in \mathbb{N})$$

Relation (3.1.11) represents the generalization of the Chapman–Kolmogorov equation (3.1.3) (with $\nu=1$ and μ replaced by $\mu-1$) for Markov renewal processes.

Furthermore, it holds that

$$(3.1.12) \qquad R_{ij} = \sum_{\mu=0}^{\infty} Q_{ij}^{(\mu)} \qquad (i,j \in \mathcal{E})$$

Relation (3.1.12) together with the fact that the functions $Q_{ij}^{(\mu)}$ are continuous from the right provide the continuity from the right of the renewal functions R_{ij}. We summarize the properties of R_{ij} in

(3.1.13) **Theorem.**
For all states $i,j \in \mathcal{E}$, the renewal functions R_{ij} are increasing and continuous from the right. For all states i and all transient states j, the functions R_{ij} are bounded.

Using (3.1.11) and (3.1.12) it can be shown that the renewal functions satisfy the system of integral equations

$$(3.1.14) \qquad R_{ij}(t) = \delta_{ij}(t) + \sum_{k \in \mathcal{E}} \int_{[0,t]} Q_{kj}(t-s) R_{ik}(ds) \qquad (i,j \in \mathcal{E}; t \geq 0)$$

3.2 STEOR Networks and Markov Renewal Processes

(3.2.1) In what follows, let $V=\{1,\ldots,n\}$ be the node set of the STEOR network N in question. For practical reasons, we add to N a dummy sink $n+1$, which is supposed to be a STEOR node, and for each sink j of N the dummy activity $<j,n+1>$ (of duration 0). Thus, arc $<j,n+1>$ ($j\in S$) has the weight $\begin{bmatrix} 1 \\ I_0 \end{bmatrix}$ where I_0 is the unit step function with jump point at 0. This yields the **expanded STEOR network** N^+ with node set $V^+:=\{1,\ldots,n,n+1\}$. In order to assign weight vectors to all node pairs (i,j) with $i,j\in V^+$, we put

$$p_{n+1,n+1}:=1,\ F_{n+1,n+1}:=I_0$$

$$p_{ij}:=0,\ F_{ij}:=I_0 \quad \text{for all remaining node pairs } (i,j)$$
$$\text{for which } N^+ \text{ contains no arc } <i,j>$$

(3.2.2) At first we consider a STEOR network N with only one source. Each realization of N begins with the activation of the single source. Every node of N represents a transit node (that is, whenever an incoming activity has been terminated, an outgoing activity is carried out). Hence, precisely one walk of the STEOR network N or the expanded STEOR network N^+, respectively, whose initial node is the source, is executed during each project realization and thus $D^{skip}=D$. The walk executed terminates either in a sink of N or, respectively, the dummy sink of N^+ (later we will see that this happens with probability 1) or in a cycle structure of N (with probability 0). In the first case, the walk has finitely many nodes (where arcs executed several times and their initial and final nodes are counted as often as their number of executions indicates). In the latter case, the walk has an infinite number of nodes (infinite cycling in a cycle structure).

(3.2.3) In either case those *nodes* of a STEOR network which are activated during a project realization *are activated one after another* (nodes inside cycles perhaps several times), and at most one sink of a STEOR network is activated. In particular, assumption A6 is satisfied and, trivially, A4 and A5 too (where again convention (1.3.11) is taken into account). Thus, each STEOR network represents an admissible GERT network.

Now let the index $\nu \geq 0$ count the successive activations of nodes of the expanded STEOR network N^+ where the zeroth node activation is the activation of the source at time 0. Let X_ν be that node whose activation is the νth node activation in turn and let θ_ν be the time of that activation ($\nu \in \mathbb{N}_0$). Then

$$0 = \theta_0 \leq \theta_1 \leq \theta_2 \leq \dots$$

Moreover, using the homogeneity and Markov properties from assumptions A2c and A2d as well as relation $P(A \cap B | C) = P(A | B \cap C) P(B | C)$, we have for all $\nu \in \mathbb{N}_0$; $i, j \in V^+$, and $t \geq 0$

$$P(X_{\nu+1} = j, \theta_{\nu+1} - \theta_\nu \leq t \,|\, X_\nu = i, X_{\nu-1}, \dots, X_0; \theta_\nu, \dots, \theta_0)$$

$$= P(X_{\nu+1} = j, \theta_{\nu+1} - \theta_\nu \leq t \,|\, X_\nu = i)$$

$$= P(\theta_{\nu+1} - \theta_\nu \leq t \,|\, X_{\nu+1} = j, X_\nu = i) \, P(X_{\nu+1} = j \,|\, X_\nu = i) = p_{ij} F_{ij}(t)$$

(compare also remarks (1.3.6c) and (1.3.6d)). Hence, we have proved

(3.2.4)   **Theorem.**

Let N be a STEOR network with only one source, say node 1, and let N^+ be the corresponding expanded network. Furthermore, let X_ν be that node of N^+ whose activation is the νth node activation in turn, and let θ_ν be the time of that node activation ($\nu \in \mathbb{N}_0$). Then the sequence $(X_\nu, \theta_\nu)_{\nu \in \mathbb{N}_0}$ represents a homogeneous Markov renewal process with state space $V^+ \times \mathbb{R}_+$ whose transition functions Q_{ij} are given by

$$Q_{ij}(t) = p_{ij} F_{ij}(t) \qquad (i, j \in V^+, \ t \in \mathbb{R})$$

and whose initial distribution by $p_1 = 1$, $p_i = 0$ for $i = 2, \dots, n+1$. $(X_\nu)_{\nu \in \mathbb{N}_0}$ is a homogeneous Markov chain with state space V^+, transition probabilities p_{ij} $(i,j \in V^+)$, and initial distribution $p_1 = 1$, $p_i = 0$ for $i = 2, \ldots, n+1$.

Next, we want to find out the type of the states of the Markov chain $(X_\nu)_{\nu \in \mathbb{N}_0}$, which correspond to the nodes of the expanded STEOR network N^+. We see immediately that the dummy sink n+1 corresponds to an absorbing state (because $p_{n+1,n+1} = 1$) and that each node $i \in V$ outside any cycle represents a transient node (the probability f_{ii}^* that such a node i is activated a second time equals 0). Now let i be a node belonging to a cycle. Owing to assumption A3, there is a path from i to a node $j \in V$ outside any cycle such that the execution probabilities of all activities of that path are positive. Hence, if that path contains μ arcs, we have $p_{ij}^{(\mu)} > 0$ (by exploiting the Chapman–Kolmogorov equation (3.1.3)). Since by theorem (3.1.4), for a recurrent state i and a transient state j, it holds that $p_{ij}^{(\mu)} = 0$ for all $\mu \in \mathbb{N}$, node i cannot be a recurrent state. Thus, we have

(3.2.5)   Proposition.
The dummy sink of the expanded STEOR network N^+ represents an absorbing state, and all remaining nodes represent transient states of the associated Markov chain.

Now we prove

(3.2.6)   Theorem.
For a STEOR network with only one source, at most one and with probability 1 exactly one terminal event occurs during each project execution. Moreover, $D^{skip} = D$ and $P(D < \infty) = 1$.

Proof.
We already know that at most one sink of a STEOR network N with only one source is activated and that $D^{skip} = D$. Now let node 1 be the source of N or N^+, respectively. Since all nodes $j \in V$ represent transient states, we have by theorem (3.1.5a)

$$\lim_{\mu \to \infty} p_{1j}^{(\mu)} = 0 \quad \text{for } j \in V$$

Because of

$$\sum_{j \in V^+} p_{1j}^{(\mu)} = 1 \quad \text{for all } \mu \in \mathbb{N}$$

(cf. (3.1.2)) it follows that

$$\lim_{\mu \to \infty} p_{1,n+1}^{(\mu)} = 1$$

that is, the dummy sink of N^+ and thus exactly one sink of N is activated with probability 1 during each project execution.

By theorem (3.1.5b) the number of times any node of N is activated and thus any activity of N is carried out is finite with probability 1. Since $P(D_{ij} < \infty) = 1$ for every activity $<i,j>$ (cf. (1.2.5)), the project duration D is finite with probability 1.

∎

(3.2.7)   **Remark.**
Theorem (3.2.6) says in particular that each STEOR network with only one source satisfies the one–sink condition. By (2.3.3) we thus have for the distribution function G of the project duration D (which coincides with the distribution function G^{skip} of the skipping project duration D^{skip})

(3.2.8) $$G = \sum_{j \in S} Y_j$$

Recall that each cycle structure of a weakly admissible GERT network contains only STEOR nodes. From theorem (3.2.6) we then obtain the following

(3.2.9)   **Corollary.**
Let C be any cycle structure of a weakly admissible GERT network with possibly several sources and let i be an entrance node of C. Then the **sojourn time in** C when C is **entered at** node i (that is, the time elapsed between an activation of i and the moment at which the execution of an exit arc of C is begun implied by that activation of i) is finite with probability 1. In particular, the number of node activations in C (starting

with the activation of node i when C is entered at i and stopping when C is "left") is finite with probability 1.

Corollary (3.2.9) and the fact that, for an admissible GERT network, the total time of carrying out all activities outside cycles is finite with probability 1 yield

(3.2.10)   Theorem.
For an admissible GERT network with possibly several sources it holds that $P(D<\infty) = 1$.

Since the one–source network that corresponds to a STEOR network with several sources (cf. (1.3.4)) does not represent a STEOR network unless all feasible subsets of R are singletons, it is expedient to discuss STEOR networks with several sources separately. Trivially, each STEOR network N with several sources is an admissible GERT network. As we will see in section 3.4, the number of Markov renewal processes associated with N equals the maximum number of sources of N that can be activated jointly during a single realization of N. If r sources are activated during a realization of N, then, because of assumptions A1 and A2a, r disjoint walks of N are executed. Each of the r walks originates at one of the r sources and terminates either in a sink of N (with probability 1) or in a cycle structure of N (with probability 0). If at least one walk terminates in a sink and at least one walk terminates in a cycle structure during a project realization ω, we have $D^{skip}(\omega)<D(\omega)$. Thus, in general it holds that $D^{skip}\leq D$ but $P(D^{skip}=D)=1$. Observing assumptions A1 and A2a we can show in analogy to the proof of theorem (3.2.6)

(3.2.11)   Theorem.
If a source i of a STEOR network N is activated during an execution of N, then at most one and with probability 1 exactly one sink $j\in\mathcal{R}(i)$ is activated. Moreover, $D^{skip}\leq D$ and $P(D^{skip}=D)=1$.

(3.2.12)   Remark.
Theorem (3.2.11) says in particular that a STEOR network with several sources satisfies the one–sink condition.

(3.2.13) As mentioned above, a single Markov renewal process can be assigned to a STEOR network N with source set R where $|R|>1$ (or to the respective expanded STEOR network N^+) if only one source is activated during any realization of N. The transition functions of that Markov renewal process are again given by $Q_{ij}(t)=p_{ij}F_{ij}(t)$ $(i,j\in V^+; t\in\mathbb{R})$. The initial distribution is given by $p_i=q_i$ for $i\in R$ (where q_i is again the probability that node i is activated) and $p_i=0$ for $i\in V^+\setminus R$. STEOR networks where more than one source can be activated during a single network realization will be discussed in section 3.4 (cf. (3.4.5)).

3.3 Basic Properties of Admissible EOR Networks

If each node of a GERT network has an EOR entrance but may have a deterministic or a stochastic exit, we speak of an **EOR network**. In what follows, we will show that if a certain condition (which can be examined very easily) is satisfied, a GERT network N without AND nodes is admissible and each of its IOR nodes is non–genuine and thus has to be replaced by an EOR node, in other words, N represents an admissible EOR network. First we state the following

(3.3.1)   Condition.
For each deterministic node i and for any two distinct successors k_1 and k_2 of i it holds that $\mathcal{R}(k_1)\cap\mathcal{R}(k_2)=\emptyset$.

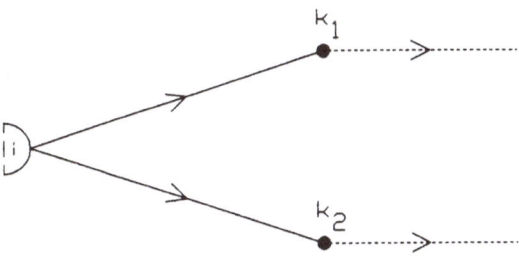

Figure 3.3.1

Condition (3.1.1) says, figuratively speaking, that different walks separating at a deterministic node i do not meet beyond node i (see Fig. 3.3.1). It can be tested in $O(|V|^2)$ time whether condition (3.3.1) is satisfied in a given GERT network (cf. Rubach (1984)). Next we prove the basic

(3.3.2)   Theorem.

Let N be a GERT network where all nodes outside cycles with more than one predecessor have IOR entrance and where all remaining nodes have EOR entrance. Exactly if N satisfies condition (3.3.1), N represents an admissible EOR network.

Proof (cf. Neumann (1985a)).

The GERT network N satisfies assumption A5 and the "entrance part" of assumption A4. The conditions to be met in addition by an admissible EOR network N are as follows:

(a) Every node of N that belongs to a cycle is stochastic ("exit part" of assumption A4).

(b) For each node of N outside any cycle, at most one of its incoming activities is carried out during a single realization of N (that is, an EOR entrance is to be assigned to such a node according to convention (1.3.11)).

(c) During each realization of N, at most one entrance arc of each cycle structure of N is executed (assumption A6).

Necessity.

It is easy to see that assuming conditions (a), (b), and (c) are fulfilled but condition (3.3.1) is violated provides a contradiction. Indeed, if condition (3.3.1) is not satisfied, there are a deterministic node i and two distinct walks W_1 and W_2 from node i to a node

j (see Fig. 3.3.2). If j does not belong to a cycle, then condition (b) is violated, if j belongs to a cycle, then condition (c) is not satisfied.

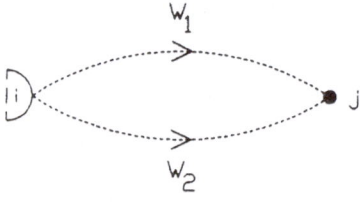

Figure 3.3.2

Sufficiency.

Trivially, if condition (a) does not hold, then condition (3.3.1) is violated (compare the

example in Fig. 3.3.3 where for the two distinct successors 3 and 4 of node 2 it holds
that $\mathcal{R}(3) \cap \mathcal{R}(4) = \mathcal{R}(4) \neq \emptyset$).

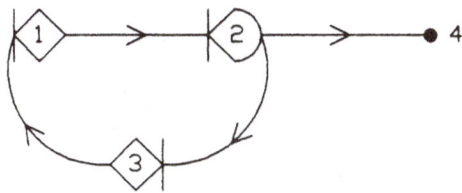

Figure 3.3.3

Now assume that condition (3.3.1) is satisfied, condition (b) is violated for a node j of
N, but (b) holds for every node before j (in other words, node j is the "first node" that
violates (b)), and (c) holds for each cycle structure before j (for the concept "before"
compare (1.1.1) and (1.1.2)). Then, by observing assumption A1, there exist a
deterministic node i and two walks W_1 and W_2 from i to j that have only the nodes i
and j in common (see Fig. 3.3.4). Thus, for the successors k_1 and k_2 of i on W_1 and W_2,
respectively, we have $j \in \mathcal{R}(k_1) \cap \mathcal{R}(k_2)$ in contradiction to condition (3.3.1).

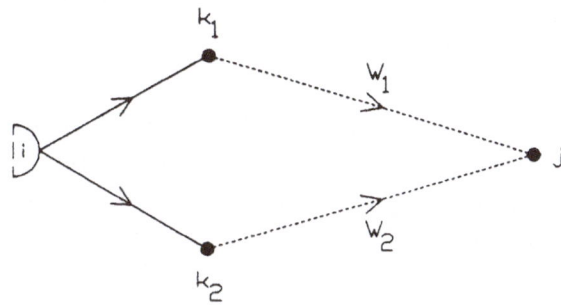

Figure 3.3.4

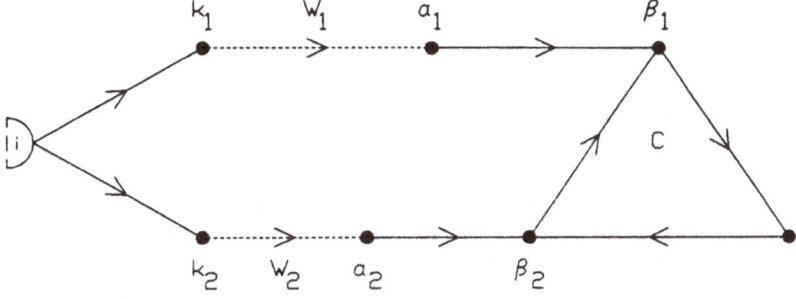

Figure 3.3.5

Next assume that (3.3.1) is satisfied, condition (c) is violated for a cycle structure C, but (c) holds for every cycle structure before C (that is, C is the "first cycle structure" that violates (c)), and (b) holds for every node before C. Let $\langle\alpha_1,\beta_1\rangle$ and $\langle\alpha_2,\beta_2\rangle$ be two distinct entrance arcs of C that are executed during the same realization of N. Then by assumption A1 there exist a deterministic node i and two walks W_1 and W_2 from i to α_1 and α_2, respectively, which have only node i in common (see Fig. 3.3.5). Let k_1 and k_2 be the successors of i on W_1 and W_2, respectively, and let j be any node of C. Then $j\in\mathcal{R}(k_1)\cap\mathcal{R}(k_2)$ contradicts condition (3.3.1).

∎

(3.3.3) Next, we will see that an admissible EOR network has an interesting tree–structure property. First we introduce two concepts. A subnetwork N' of a GERT network N is said to **correspond to a realization** ω of N if the arc set of N' coincides with the set of those activities from N which are carried out during project realization ω. A cycle structure C of an admissible GERT network N is said to be **shrunk** to one STEOR node k if C is replaced by k, where all entrance arcs and exit arcs of C form the incoming arcs and outgoing arcs, respectively, of node k (the arc weights are of no interest in this context). Since at most one entrance arc of C is carried out during any realization of N and the execution of an entrance arc induces the execution of exactly one exit arc with probability 1, the "substitute" node k really behaves like a STEOR node (with probability 1). Then the following **tree–structure property** is immediate from condition (3.3.1) and assumption A1:

(3.3.4)   **Proposition.**
If we shrink each cycle structure of an admissible EOR network N to one STEOR node, then any subnetwork of N that corresponds to a single realization of N represents an outforest [6].

---

6   For the concept outforest compare (1.1.3).

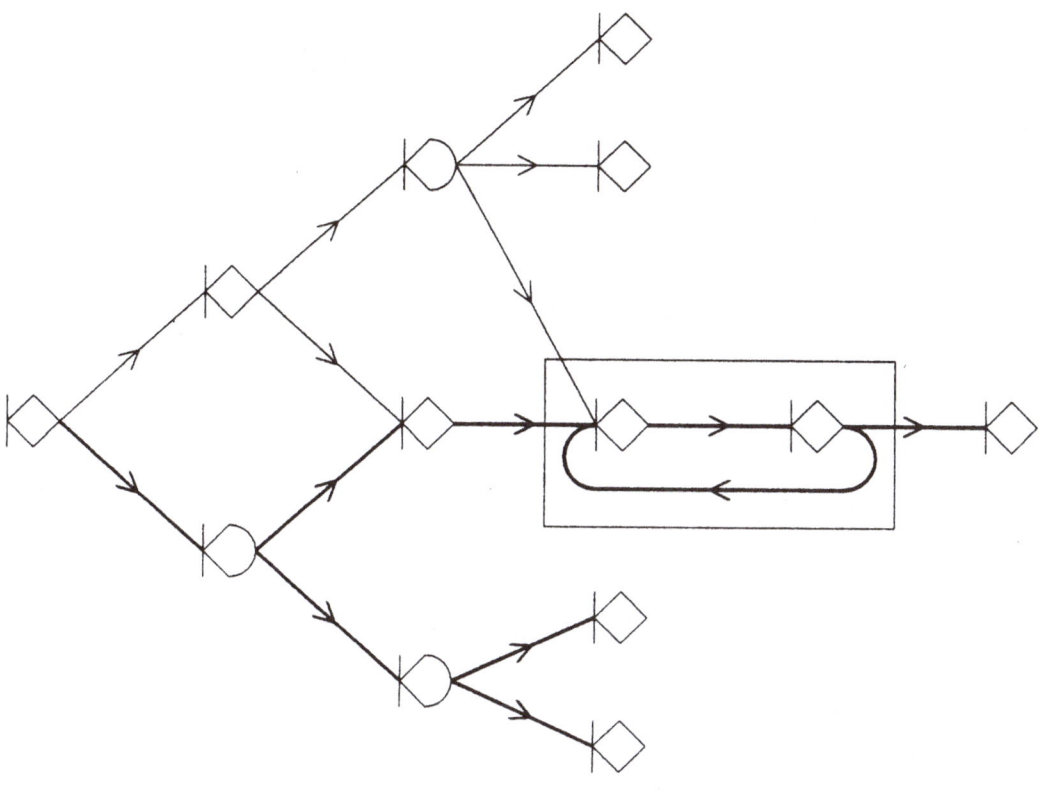

Figure 3.3.6

Note that even if all cycle structures of an admissible EOR network N are shrunk to single nodes, N itself does not form a directed forest; only a subnetwork that corresponds to a single project realization represents, after the shrinkage of cycle structures, an outforest. In Fig. 3.3.6 the darker subnetwork forms an outtree after shrinking the cycle inside the frame to a STEOR node.

3.4 Coverings of Admissible EOR Networks

(3.4.1) In what follows we show that each admissible EOR network can be "covered" by several STEOR networks. For this we first introduce the concept of the Markov degree of an admissible EOR network N. Let $d(\omega)$ be the number of sinks of N that are activated during network realization ω. Then

$$d := \max_{\omega \in \Omega} d(\omega)$$

is called the **Markov degree** of N, where Ω is again the set of all possible realizations of N. Later we will see that an admissible EOR network with Markov degree d can be associated with d Markov renewal processes. By theorem (3.2.6) we have d=1 for a STEOR network with only one source. If the EOR network contains at least one deterministic node, then d>1.

For an admissible GERT network N, at most one entrance arc of each cycle structure C of N is carried out during each realization of N. With probability 1, the execution of an entrance arc of C implies the execution of exactly one exit arc of C, whereas with probability 0, cycle structure C is not left. Thus, the shrinkage of cycle structures explained in (3.3.3) does not alter the Markov degree of an admissible EOR network. This fact is stated as

(3.4.2) Proposition.
If we shrink each cycle structure of an admissible EOR network N to one STEOR node, the Markov degree of N remains unchanged.

To compute the Markov degree of an admissible EOR network N, we introduce the concept of the Markov degree of a node of N and state two theorems. Let i be a node of N which does not belong to any cycle and is different from a sink, and let N(i) be the GERT subnetwork of N induced by $\mathcal{R}(i)$ (cf. (1.4.5)) treated as a separate entity. Then the **Markov degree** d_i **of node** i is defined to be the Markov degree of the EOR network N(i). For a sink i we put $d_i := 1$. The Markov degree of any node belonging to a cycle structure C is to be equal to the Markov degree of the "substitute" STEOR node j obtained by shrinking C to node j.

(3.4.3) Theorem.
Let i be a node of an admissible EOR network different from a sink. Then

$$d_i = \begin{cases} \max_{j \in \mathcal{S}(i)} d_j & \text{if i is stochastic} \\ \sum_{j \in \mathcal{S}(i)} d_j & \text{if i is deterministic} \end{cases}$$

Proof.

We may restrict ourselves to nodes outside any cycle because all (stochastic) nodes belonging to one and the same cycle structure possess the same Markov degree.

Let i be a stochastic node outside any cycle and different from a sink. Then exactly one arc emanating from i is carried out when i has been activated. Hence, the maximum number of sinks activated during any realization of $N(i)$ equals the maximum of the Markov degrees of the successors of i.

If node i is deterministic, then all arcs emanating from i are carried out and thus all successors of i are activated when i has been activated. By theorem (3.3.2), for any two distinct successors j_1 and j_2 of i, it holds that $\mathcal{R}(j_1) \cap \mathcal{R}(j_2) = \emptyset$. Hence, the sets of sinks activated during any realization of $N(j_1)$ on the one hand and of $N(j_2)$ on the other hand are disjoint. Moreover, by assumption A2, the executions of networks $N(j_1)$ and $N(j_2)$ are independent. As a consequence, the maximum number of sinks activated during any realization of $N(i)$ equals the sum of the Markov degrees of the successors of i.

∎

(3.4.4)   Theorem.

Let $\hat{P}(R)$ be the set of all feasible subsets of the source set R of an admissible EOR network N (see remark (1.3.2a)). Then the Markov degree of N is

$$d = \max_{R' \in \hat{P}(R)} \sum_{i \in R'} d_i$$

Proof.

If N has only one source, say node 1, then $d = d_1$. For $|R| > 1$, let $\tilde{N}$ be the one–source network corresponding to N introduced in (1.3.4), which also represents an admissible EOR network. Let i_0 be the source and let the nodes $i_{R'}$ $(R' \in \hat{P}(R))$ be the successors of the source i_0 in $\tilde{N}$. Obviously, the Markov degrees of N and $\tilde{N}$ coincide and thus the Markov degree of N equals the Markov degree d_{i_0} of source i_0 of $\tilde{N}$. By theorem

(3.4.3) $d_{i_0} = \max\limits_{R' \in \mathbb{P}(R)} d_{i_{R'}}$. If $|R'|=1$, say $R'=\{i'\}$, then $d_{i_{R'}} = d_{i'}$. If $|R'|>1$, then by

theorem (3.4.3) $d_{i_{R'}} = \sum\limits_{i \in R'} d_i$. Hence $d_{i_0} = \max\limits_{R' \in \mathbb{P}(R)} \sum\limits_{i \in R'} d_i$.

∎

(3.4.5) For a STEOR network with several sources, the Markov degree d equals the maximum number of sources that can be activated jointly during a single project realization, that is,

$$d = \max_{R' \in \mathbb{P}(R)} |R'| \leq \min(|R|, |S|)$$

Owing to proposition (3.4.2) and theorems (3.4.3) and (3.4.4), to find the Markov degree of an admissible EOR network N we may restrict ourselves to computing the Markov degrees d_i of the sources i of N where N can be assumed to be acyclic. Since the nodes of an acyclic network can be ordered topologically such that $j>i$ for $j \in \mathcal{S}(i)$ (cf. (1.1.1)), the Markov degrees of the nodes of N can be computed "backwards" by means of the formulas from theorem (3.4.3) beginning with $d_j=1$ for each sink j.

Next, we introduce the concept of a covering of an admissible EOR network.

(3.4.6)   Definition.

A set of k different subnetworks $N_1, \ldots, N_k$ of an admissible EOR network N, where N_κ has node set V_κ ($\kappa=1, \ldots, k$), is called a **covering** of N of **size** k if for $\kappa=1, \ldots, k$

(a)   each source of N_κ is a source of N

(b)   when $i \in V_\kappa$ is deterministic, then exactly one outgoing arc belongs to N_κ, and when $i \in V_\kappa$ is stochastic, then all outgoing arcs belong to N_κ

(c)   each walk from a source to a sink in N belongs to at least one subnetwork N_κ

(d)   at most one source of N_κ is activated during any realization of N.

If $\{N_1, \ldots, N_k\}$ is a covering of N, we also say that the networks $N_1, \ldots, N_k$ **cover** N.

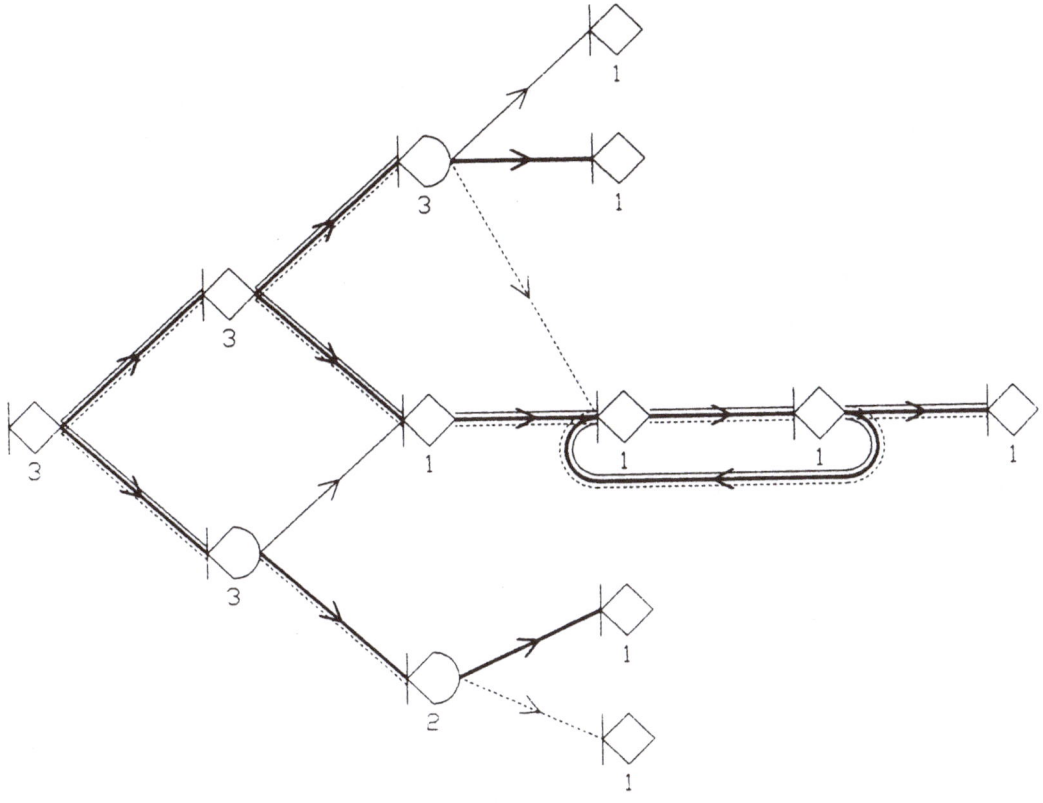

Figure 3.4.1

The admissible EOR network of Fig. 3.4.1 has Markov degree 3 (the nodes in Fig. 3.4.1 are marked with their Markov degrees) and possesses a covering of size 3. The arcs of the three subnetworks of that covering are illustrated by normal, bold, and dotted arrows. Now we prove

(3.4.7)   Theorem.

The minimum size of a covering of an admissible EOR network N is equal to the Markov degree of N.

Proof.

Owing to proposition (3.4.2) we may assume without loss of generality that N is acyclic.

At first we consider the case where N has only one source. Then condition (3.4.6d) is satisfied automatically.

Let d be again the Markov degree of N. At first we prove that there is a covering of N of size d. Since N is acyclic, the nodes of N can be topologically ordered such that $j>i$ for each $j \in \mathcal{S}(i)$. Then we construct a covering of $N(i)$ given coverings of all $N(j)$. In more detail, let $i \in V \backslash S$. We show that given d_j subnetworks $N_1(j), \ldots, N_{d_j}(j)$ of $N(j)$ for each $j \in \mathcal{S}(i)$ [7] which satisfy the conditions (a), (b) and (c) of definition (3.4.6) (with N replaced by $N(j)$), there can be constructed d_i subnetworks $N_1(i), \ldots N_{d_i}(i)$ of $N(i)$ that also satisfy those conditions (with N replaced by $N(i)$).

Let $i \in V \backslash S$ be a stochastic node and thus $d_i = \max_{j \in \mathcal{S}(i)} d_j$. For each $j \in \mathcal{S}(i) \backslash S$ with $d_j < d_i$, we introduce $d_i - d_j$ additional subnetworks of $N(j)$ by $N_{d_j+1}(j) := \ldots := N_{d_i}(j)$ $:= N_1(j)$. Then for $\delta = 1, \ldots, d_i$, the subnetworks $N_\delta(i)$ of $N(i)$ are given by defining their arc sets $E_\delta(i)$ as

$$E_\delta(i) := (\bigcup_{j \in \mathcal{S}(i)} \{<i,j>\}) \cup (\bigcup_{j \in \mathcal{S}(i) \backslash S} E_\delta(j))$$

where $E_\delta(j)$ is the arc set of $N_\delta(j)$. It is easy to see that the subnetworks $N_1(i), \ldots, N_{d_i}(i)$ of $N(i)$ satisfy the conditions (a), (b) and (c) of definition (3.4.6) with $N(i)$ instead of N.

Now let $i \in V \backslash S$ be a deterministic node and thus $d_i = \sum_{j \in \mathcal{S}(i)} d_j$. For each $j \in \mathcal{S}(i) \cap S$ we have $d_j = 1$, and the arc $<i,j>$ together with its initial and final nodes form a subnetwork of $N(i)$. For each $j \in \mathcal{S}(i) \backslash S$, the d_j networks with arc sets $\{<i,j>\} \cup E_\delta(j)$ ($\delta = 1, \ldots, d_j$) represent subnetworks of $N(i)$, where $E_\delta(j)$ is again the arc set of $N_\delta(j)$.

---

[7]   For simplicity, if $j \in S$, $N(j)$ consisting only of the isolated node j is also considered to be a subnetwork.

The d_i subnetworks of $N(i)$ constructed in this way satisfy the conditions (a), (b) and (c) of definition (3.4.6) with $N(i)$ instead of N.

Secondly we show that there is no covering of N with size k<d. Assume that $\{N_1, \ldots, N_k\}$ represents a covering of N. Since d is the Markov degree of N, there is a network realization during which d paths from the source to d distinct sinks of N are executed. By condition (3.4.6c), each such path belongs to at least one of the subnetworks $N_1, \ldots, N_k$. Because of k<d at least one subnetwork, say N_κ, contains two of the paths, say W_1 and W_2. The paths W_1 and W_2 are disjoint beyond some deterministic node, that is, N_κ contains a deterministic node i and two distinct successors j_1 and j_2 of i, where $\langle i, j_1 \rangle$ belongs to W_1 and $\langle i, j_2 \rangle$ belongs to W_2, in contradiction to condition (3.4.6b).

If the EOR network N in question has more than one source, we consider the one–source network $\tilde{N}$ corresponding to N. Obviously, each "restriction" of a covering $\tilde{C}$ of $\tilde{N}$ to N satisfies condition (3.4.6d) and provides a covering C of N where $\tilde{C}$ and C have the same size. On the other hand, there may be a covering of N which does not represent a restriction of a covering of $\tilde{N}$, however, there is no covering of N of size less than d.

The latter case occurs if N contains two sources i and j whose Markov degrees are different, say $d_i > d_j$. Then in each covering $\tilde{C}$ of $\tilde{N}$ of minimum size d and thus in the restriction C' of $\tilde{C}$ to N, both nodes i and j belong to d_i covering subnetworks. On the other hand, there is a covering C of N such that source i belongs to d_i and source j belongs to d_j covering subnetworks. However, C' can be "reduced" to a covering of "type" C by omitting $d_i - d_j$ appropriate walks with initial node j from the covering networks, where the size of the covering is conserved. This completes the proof. ∎

(3.4.8)   Remark.
In general, there is more than one covering of size d for an admissible EOR network with Markov degree d.

An example of the case discussed at the end of the proof of theorem (3.4.7) is illustrated in Figs. 3.4.2 and 3.4.3. Fig. 3.4.2 shows an admissible EOR network N (without arc weights) with Markov degree 2. Because of assumption A1 exactly one of the two sources of N is activated during each network realization. A covering C of N of size 2 is illustrated by normal and darker arrows in Fig. 3.4.2. Fig. 3.4.3 shows the one–source network Ñ corresponding to N and a covering $\tilde{C}$ of Ñ of size 2. In the restriction C' of $\tilde{C}$ to N, arc <2,4> and nodes 2 and 4 belong to both covering networks. If we eliminate arc <2,4> and nodes 2 and 4 from the "darker subnetwork" of C', we obtain the covering C.

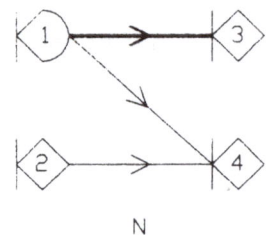

N

Figure 3.4.2

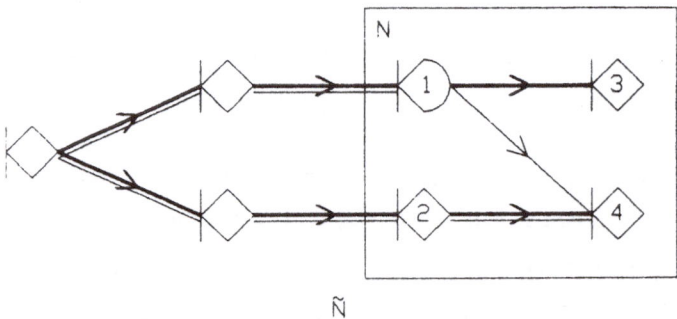

Ñ

Figure 3.4.3

Each deterministic node in a covering network N_κ has exactly one successor and thus has to be replaced by a stochastic node in accordance with convention (1.2.23) if N_κ is treated as a separate entity. Hence, each covering network N_κ, if taken on its own, represents a STEOR network and, owing to (3.4.5) and condition (3.4.6d), has Markov degree 1. As a consequence, each network N_κ can be associated with a single Markov renewal process (cf. (3.2.13)). Thus, we have

(3.4.9) Proposition.
An admissible EOR network N with Markov degree d can be covered by d STEOR networks each of which has Markov degree 1. Thus, N can be associated with d Markov renewal processes.

If r sources are activated during a realization of an admissible EOR network N with Markov degree d (where $r \leq d$), then at least r and at most d different walks of N are executed. Every two different such walks are either completely disjoint (if they originate at different sources of N) or disjoint beyond some deterministic node (if they originate at the same source). Each walk terminates either in a sink of N (with probability 1) or in a cycle structure (with probability 0).

In analogy to the basic theorem (3.2.11) for STEOR networks we have

(3.4.10) Theorem.
If a source i (with Markov degree d_i) of an admissible EOR network N is activated during an execution of N, then at most d_i sinks and with probability 1 at least one sink from $\mathcal{R}(i)$ are activated. Moreover, $D^{skip} \leq D$ and $P(D^{skip} = D) = 1$.

Since it may happen that all realizable sinks have been activated and some activities of walks terminating in a cycle structure are still in execution, D may be greater than D^{skip} even for an admissible EOR network with only one source. The probability of this random event, however, equals 0.

(3.4.11) It is immediate from definition (3.4.6) that, for each admissible EOR network N, there exists a covering $\{N_1, \ldots, N_k\}$ of N such that for each node j of N, there is a subnetwork N_κ ($1 \leq \kappa \leq k$) which contains all nodes from $\mathcal{R}(j)$ [8] and all arcs of N joining those nodes. The size k of this covering, however, may be greater than the Markov degree d of N. As an example, we consider the admissible EOR network N of Markov degree 2 in Fig. 3.4.4 (where the nodes are again marked with their Markov degrees).

---

8   The set $\mathcal{R}(j)$ refers to the network N.

The two subnetworks illustrated by normal and darker arrows form a covering of N. None of the two subnetworks contains all nodes from $\mathbb{Z}(5)=\{1,2,4,5\}$ and all joining arcs $<1,2>,<1,4>,<2,5>$, and $<4,5>$. The subnetwork shown by dotted arrows contains all nodes from $\mathbb{Z}(5)$ and all arcs joining those nodes. If we add the latter subnetwork to the covering, we obtain a covering of size 3 with the desired property.

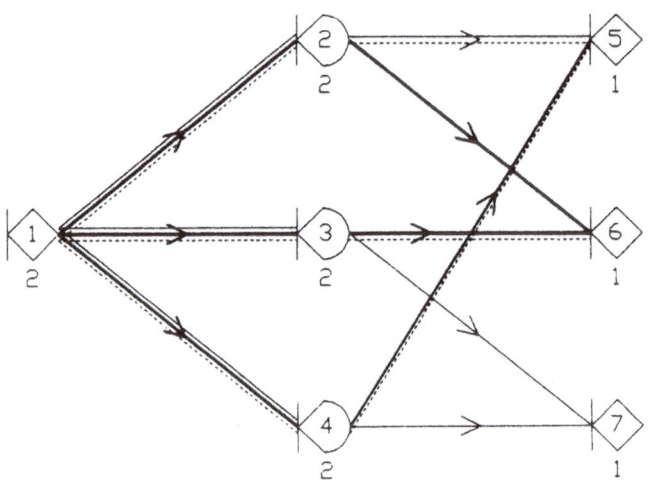

Figure 3.4.4

3.5 Properties and Computation of Activation Functions and Activation Numbers

At first we deal with **STEOR networks with only one source.** Let $V=\{1,\ldots,n\}$ be again the node set of the underlying STEOR network N and let node 1 be the source. We consider the Markov renewal process with renewal functions R_{ij} assigned to the corresponding expanded STEOR network N^+. Recall that the function value $Y_j(t)$ of the activation function of node j is the expected number of activations of node j by time t during a project execution. $Y_j(t)$ equals the expected number of "renewals" of state j of the associated Markov renewal process during the time interval $[0,t]$, where the process starts at state 1 (the source) at time 0. In other words, $Y_j=R_{1j}$ for $j=1,\ldots,n$. Since all states $j=1,\ldots,n$ are transient (cf. proposition (3.2.5)), we have by theorem (3.1.13)

(3.5.1) Proposition.

For each node j of a STEOR network with only one source, the activation function Y_j is increasing, bounded, and continuous from the right, and the activation number z_j exists (and is finite).

(3.5.2) Remarks.

(a) To show the boundedness of the activation functions, the fact stated in corollary (3.2.9) is exploited that, for any cycle structure C and any entrance node i of C, the sojourn time in C when C is entered at node i is finite with probability 1 (compare Cinlar (1975), sections 9.1 and 10.2).

(b) For an arbitrary admissible GERT network with only one source, the activation functions Y_j of nodes j in cycle structures behave like activation functions in "corresponding" STEOR networks. For nodes j outside cycles, the functions Y_j are bounded by 1 as well as increasing and continuous from the right by definition (for the continuity from the right cf. Neumann and Steinhardt (1979a), section 3.2.1). Hence, proposition (3.5.1) is also true for admissible GERT networks with only one source.

Now we return to STEOR networks. Recall the system of integral equations (3.1.14) for the renewal functions

$$R_{ij}(t) = \delta_{ij}(t) + \sum_{k \in \mathcal{E}} \int_{[0,t]} q_{kj}(t-s) R_{ik}(ds) \qquad (i,j \in \mathcal{E}; t \geq 0)$$

where

$$\delta_{ij}(t) = \begin{cases} 1, & \text{for } i=j \text{ and } t \geq 0 \\ 0, & \text{otherwise} \end{cases}$$

$$q_{kj}(t) = p_{kj} F_{kj}(t) \qquad \text{for } t \in \mathbb{R}$$

We then obtain the following system of integral equations for the activation functions of a STEOR network with only one source:

(3.5.3)
$$
\begin{cases}
Y_1(t) = 1 \\
Y_j(t) = p_{1j}F_{1j}(t) + \sum_{k=2}^{n} p_{kj} \int_{[0,t]} F_{kj}(t-s)Y_k(ds), \quad j=2,\ldots,n
\end{cases} \quad (t \geq 0)
$$

Note that $p_{kj}=0$ for $k \notin \mathcal{P}(j)$. It can be shown that the system of integral equations (3.5.3) has a unique solution (cf. Nicolai (1980)). The activation numbers satisfy the system.

(3.5.4)
$$
\begin{cases}
z_1 = 1 \\
z_j = p_{1j} + \sum_{k=2}^{n} p_{kj} z_k, \quad j=2,\ldots,n
\end{cases}
$$

The computation of the activation functions Y_j ($j=1,\ldots,n$) by solving the system (3.5.3) is referred to as the **MRP method** because it is based on the fact that a Markov renewal process can be assigned to a STEOR network. Since a STEOR network satisfies the one–sink condition, the activation functions Y_j ($j \in S$) provide the complete evaluation of the network (cf. remark (2.2.16a)). In section 3.6 we will discuss the MRP method in more detail.

As an example, let us consider the STEOR network illustrated in Fig. 3.5.1, where each arc is marked with the execution probability of the corresponding activity if it is less than 1 (the distribution functions of the activity durations are omitted).

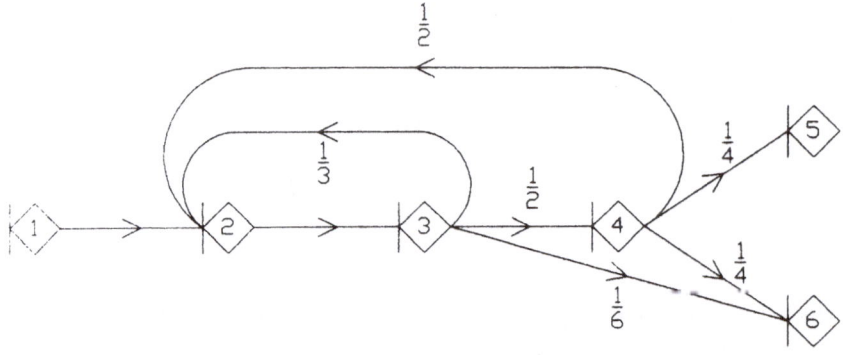

Figure 3.5.1

We seek to determine the activation probabilities q_5 and q_6 of the sinks 5 and 6, respectively. The system of equations for the activation numbers (3.5.4) takes the form

$$z_1 = 1$$
$$z_2 = 1 + \tfrac{1}{3}z_3 + \tfrac{1}{2}z_4$$
$$z_3 = z_2$$
$$z_4 = \tfrac{1}{2}z_3$$
$$z_5 = \tfrac{1}{4}z_4$$
$$z_6 = \tfrac{1}{6}z_3 + \tfrac{1}{4}z_4$$

We obtain $z_2 = z_3 = 2.4$, $z_4 = 1.2$, $z_5 = 0.3$, $z_6 = 0.7$. By (2.1.5) we get

$$q_5 = z_5 = 0.3, \quad q_6 = z_6 = 0.7$$

Sometimes we are only interested in some of the **moments of the activation functions** Y_j. Let

$$(3.5.5) \qquad \eta_j := \int_0^\infty t Y_j(dt) \qquad (j = 1, \ldots, n)$$

be the first moment of Y_j. Then

$$\mathbb{E}(D \mid A_j) = \frac{\eta_j}{q_j} \quad (j \in S)$$

(where $q_j = z_j$) is the conditional expected project duration given that sink j is activated (provided that $z_j > 0$). Moreover,

$$\mathbb{E}(D) = \sum_{j \in S} \eta_j$$

is the (unconditional) expected project duration.

The moments η_j can be found by solving a system of linear equations. Let

$$(3.5.6) \qquad d_{kj} := \int_0^\infty tF_{kj}(dt) \qquad (<k,j> \in E)$$

be the expected duration of activity $<k,j>$ and let H_{kj} be the convolution of the functions F_{kj} and Y_k, that is,

$$(3.5.7) \qquad H_{kj}(t) = \int_{[0,t]} F_{kj}(t-s)Y_k(ds) \qquad \text{for } t \geq 0$$

Then, by using (3.5.5), (3.5.6), (3.5.7) and

$$z_k = \int_0^\infty Y_k(ds)$$

we have

$$(3.5.8) \qquad \int_0^\infty tH_{kj}(dt) = \int_0^\infty tY_k(dt) + \int_0^\infty tF_{kj}(dt) \int_0^\infty Y_k(ds) = \eta_k + d_{kj}z_k$$

If we substitute in (3.5.5) the equations (3.5.3) and take into account (3.5.6), (3.5.7), and (3.5.8), we obtain the system of linear equations

$$(3.5.9) \qquad \begin{cases} \eta_1 = 0 \\ \eta_j = p_{1j}d_{1j} + \sum_{k=2}^{n} p_{kj}\eta_k + \sum_{k=2}^{n} p_{kj}d_{kj}z_k \qquad (j=2,\dots,n) \end{cases}$$

We see that for computing the moments η_j, the system of equations (3.5.4) has to be solved before.

In the same way, systems of equations for higher moments of the activation functions Y_j can be derived (compare Neumann and Steinhardt (1979a), section 3.2.2).

(3.5.10) Next, we turn to **admissible EOR networks with only one source.** Let N be such a network and j be any node of N. By (3.4.11) there is a covering of N and a STEOR network N_κ from this covering such that N_κ contains all nodes from $\mathbf{Z}(j)$ [9] and all arcs of N joining those nodes. Each deterministic node from $\mathbf{Z}(j)$ is turned into a stochastic node in N_κ. Any such node has exactly one outgoing activity in N_κ whose execution probability of 1 is the same as in N. Hence, the activation function Y_j of j in N coincides with the activation function Y_j^κ of j in N_κ. As a consequence, the activation functions of an admissible EOR network also satisfy the system of integral equations (3.5.3), and the activation numbers satisfy the system of equations (3.5.4).

Since an admissible EOR network N with at least one deterministic node does not satisfy the one–sink condition, the determination of the activation functions Y_j $(j \in S)$ furnishes the evaluation of N but not its complete evaluation (for a method of completely evaluating N we refer to Neumann (1985a)).

Now we take a look at **STEOR networks with several sources.** Let R be again the source set of the network. In analogy to (2.2.11) we define the **activation function** Y_{ij} of node j **given source** i by

$$(3.5.11) \qquad Y_{ij}(t) := \begin{cases} \mathbb{E}[K_j(t) \mid T_i = 0] & \text{for } t \geq 0 \\ 0 & \text{for } t < 0 \end{cases} \qquad (i \in R, j \in \mathbf{Z}(i))$$

where $\mathbb{E}[K \mid A]$ again means the conditional expectation of random variable K given the random event A. The corresponding **activation number** z_{ij} is defined as

$$(3.5.12) \qquad z_{ij} := \lim_{t \to \infty} Y_{ij}(t) \qquad (i \in R, j \in \mathbf{Z}(i))$$

---

9   The set $\mathbf{Z}(j)$ refers to the network N.

For fixed $i \in R$, the activation functions Y_{ij} ($j \in \mathcal{R}(i)$) coincide with the renewal functions R_{ij} of the Markov renewal process associated with the STEOR subnetwork induced by $\mathcal{R}(i)$ (where we have taken into account assumptions A1 and A2a) and are increasing, bounded, and right continuous, and the activation numbers z_{ij} exist.

Again, the latter fact holds to be true for every admissible GERT network. The system of integral equations (3.1.14) for the renewal functions gives the following system for the functions Y_{ij}:

$$(3.5.13) \quad \left\{ \begin{array}{l} Y_{ii}(t)=1 \quad (i \in R) \\[2mm] Y_{ij}(t)=p_{ij}F_{ij}(t)+\sum_{k \in \mathcal{R}(i)} p_{kj} \int_{[0,t]} F_{kj}(t-s)Y_{ik}(ds) \quad (i \in R, j \in \mathcal{R}(i)) \end{array} \right\} \quad (t \geq 0)$$

Recall that $p_{kj}=0$ for $k \notin P(j)$. The activation numbers z_{ij} satisfy the system

$$(3.5.14) \quad \left\{ \begin{array}{l} z_{ii} = 1 \quad (i \in R) \\[2mm] z_{ij} = p_{ij} + \sum_{k \in \mathcal{R}(i)} p_{kj} z_{ik} \quad (i \in R, j \in \mathcal{R}(i)) \end{array} \right.$$

(3.5.15) Remark.

The activation function Y_{ij} and activation number z_{ij} can also be defined for any node pair $(i,j), i \in V, j \in \mathcal{R}(i)$, to be the activation function Y_j and activation number z_j, respectively, of node j in the GERT subnetwork $N(i)$ induced by $\mathcal{R}(i)$, where $N(i)$ is treated as a separate entity. Trivially, these quantities again satisfy the systems of equations (3.5.13) and (3.5.14), respectively.

(3.5.16) The activation functions Y_{ij} ($i \in R, j \in S \cap \mathcal{R}(i)$) provide the complete evaluation of the underlying STEOR network N with source set R (cf. remark (2.2.16a)). This complete evaluation corresponds to the "reduction" of N to a bipartite STEOR network $\hat{N}$ with the same source set R and same sink set S as N and the arcs $\langle i,j \rangle$

$(i \in R, j \in Sn\mathcal{R}(i))$ with weights $\begin{bmatrix} z_{ij} \\ Y_{ij}/z_{ij} \end{bmatrix}$ [10]. The two STEOR networks N and $\hat{N}$ are

equivalent (cf. definition (2.2.17)).

For **admissible EOR networks with several sources**, we can also introduce the activation functions Y_{ij} and activation numbers z_{ij} by (3.5.11) and (3.5.12), which again satisfy the systems (3.4.13) and (3.5.14), respectively.

(3.5.17) Next, we want to compute the activation probabilities q_j of the nodes j of an admissible EOR network N (with only one source). Recall that if node j does not belong to any cycle, then $q_j = z_j$. If node j lies in a cycle, we remove all arcs with initial node j from N. In the resulting admissible EOR network N', node j represents a sink. Obviously, the activation probability q_j of node j in N equals the activation number z'_j of node j in N' and can thus be found by solving a system of linear equations of type (3.5.4). Analogously, if N has several sources, we can compute the conditional probability q_{ij} that node j is activated given that source i has been activated.

As an example, we return to the STEOR network of Fig. 3.5.1 and want to determine the activation probabilities q_2, q_3 and q_4. Trivially, $q_2 = q_3 = 1$. To find q_4 we consider the STEOR network N' obtained from network N of Fig. 3.5.1 by eliminating the arcs $\langle 4, 2 \rangle$, $\langle 4, 5 \rangle$ and $\langle 4, 6 \rangle$. Network N' is shown in Fig. 3.5.2. For the activation numbers in N' we obtain

$$z'_1 = 1$$
$$z'_2 = 1 + \frac{1}{3}z'_3$$
$$z'_3 = z'_2$$
$$z'_4 = \frac{1}{2}z'_3$$

---

[10]   For $z_{ij} = 0$ we may choose any distribution function for the duration of activity $\langle i, j \rangle$, for example, the unit step function I_0.

which provides $z_4' = q_4 = \frac{3}{4}$.

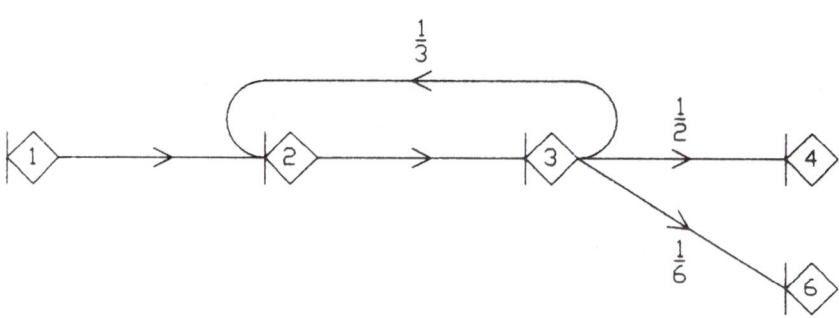

Figure 3.5.2

(3.5.18) The system of linear equations (3.5.4) for the activation numbers z_j can be solved in $O(|V|^3)$ time, for example, by the method of Gaussian elimination (cf. Stoer and Bulirsch (1980), section 4.1). If the underlying EOR network is acyclic, its nodes can be ordered topologically, which takes $O(|E|)$ time (compare (1.1.1)). Then node 1 is the source and $\mathcal{P}(j) \subseteq \{1, \ldots, j-1\}$ for each node j. In the latter case, the activation numbers $z_1, z_2, \ldots, z_n$ can be determined successively, that is, the system of linear equations is solved by "back–substitution", which requires $O(|V|^2)$ time. To find the activation probabilities q_j if the EOR network contains cycles, a system of linear equations has to be solved for each node j belonging to a cycle. This can be done in $O(|V|^4)$ time in the worst case. If the EOR network in question has several sources, the above time complexities refer to finding the quantities z_{ij} and q_{ij} for a single source i.

Sometimes, in addition to the quantities considered in section 2.3 which are related to the sink set S of the underlying admissible EOR network, we are interested in further information important to the supervision of the execution of the project. For example, we can determine the conditional probability $r_{ij}(t)$ that the terminal event j occurs not later than at a prescribed time T given that project event i has occurred at time t as follows:

$$r_{ij}(t) = Y_{ij}(T-t)$$

If the project in question can be modelled by a STEOR network, the conditional probability $r_i(t)$ that a prescribed termination time T of the project is maintained given that event i has occurred at time t is

$$r_i(t) = \sum_{j \in S} Y_{ij}(T-t)$$

If $r_i(t)$ is small, say less than a given marginal probability, it might be reasonable to discontinue the project already at time t.

At least we briefly consider the case where the **EOR network is only weakly admissible** and thus some nodes outside cycles may be activated several times during a single project execution. For weakly admissible EOR networks, the definitions (3.5.11) and (3.5.12) of the activation functions Y_{ij} and activation numbers z_{ij}, respectively, remain valid. Trivially, the functions Y_{ij} are increasing and right continuous. Each time a cycle structure C is entered, the number of node activations in C is finite with probability 1 (cf. corollary (3.2.9)). The number of activations of any node outside cycles and the number of times any cycle structure is entered are finite. Thus, the activation functions Y_{ij} are again bounded and the activation numbers z_{ij} exist (and are finite). To derive the Chapman–Kolmogorov equation (3.1.3) it is not necessary that, for fixed i and μ, the probabilities $p_{ij}^{(\mu)}$ ($j \in \mathcal{E}$) sum to unity. Similarly, relations (3.1.11), (3.1.12), and (3.1.14) hold to be true, where the boundedness of the renewal functions R_{ij} or activation functions Y_{ij}, respectively, is exploited. Hence, the systems of equations (3.5.13) and (3.5.14) for the Y_{ij} and z_{ij}, respectively, also apply to weakly admissible EOR networks.

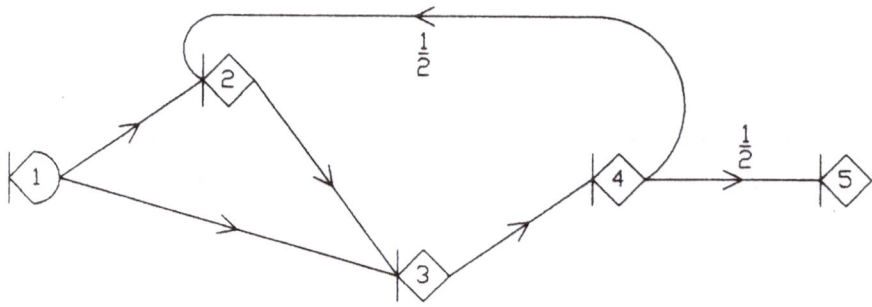

Figure 3.5.3

As an example, we consider the simple weakly admissible EOR network illustrated in Fig. 3.5.3, where each activity is marked with its execution probability if it is less than 1 (the distribution functions of the activity durations are omitted). The system of equations for the activation numbers z_j gets the form

$$z_1 = 1$$
$$z_2 = 1 + \frac{1}{2}z_4$$
$$z_3 = 1 + z_2$$
$$z_4 = z_3$$
$$z_5 = \frac{1}{2}z_4$$

We obtain $z_1 = 1$, $z_2 = 3$, $z_3 = z_4 = 4$, $z_5 = 2$. Thus, sink 5 is activated twice on the average during a single project execution.

3.6 The MRP Method

Let N be a STEOR network or a weakly admissible EOR network possibly with several sources. Recall that the MRP method consists of computing the activation functions Y_{ij} ($i \in R, j \in \mathcal{Z}(i)$) of N by solving the system of integral equations (3.5.13). Since $p_{kj} = 0$ for $k \notin \mathcal{P}(j)$, (3.5.13) and (3.5.14) become

(3.6.1) $Y_{ij}(t) = p_{ij}F_{ij}(t) + \sum\limits_{k \in \mathcal{R}(i) \cap \mathcal{P}(j)} p_{kj} \int\limits_{[0,t]} F_{kj}(t-s)Y_{ik}(ds)$

$$(i \in R, j \in \mathcal{R}(i), t \geq 0)$$

(3.6.2) $z_{ij} = p_{ij} + \sum\limits_{k \in \mathcal{R}(i) \cap \mathcal{P}(j)} p_{kj}z_{ik}$ $(i \in R, j \in \mathcal{R}(i))$

where we have omitted the trivial quantities Y_{ii} and z_{ii} ($i \in R$). If N is a STEOR network, the functions Y_{ij} provide the complete evaluation of N.

If we directly solve the system (3.6.1) of $\sum\limits_{i \in R} |\mathcal{R}(i)|$ integral equations (approximately), we speak of **version I** of the MRP method. In practice, the distributions of the activity durations and thus the distribution functions F_{kj} can rarely be estimated accurately.

Therefore, to evaluate the integrals in (3.6.1) numerically a subdivision of $[0,t]$ into subintervals, say each of length h>0, and applying the simple rectangular rule to each subinterval turn out to be expedient in most cases. The rectangular rule for a Stieltjes integral of a function F with respect to a function G over a closed interval $[a,b]$, where F and G are increasing and right continuous, has the form

$$\int\limits_{[a,b]} F(s)G(ds) \approx F(a)[G(a)-G(a-)]^{11} + F\left[\frac{a+b}{2}\right][G(b)-G(a)]$$

If, for fixed $i \in R$, we wish to evaluate the functions Y_{ij} at successive points in time $t_0=0, t_1=h, \ldots, t_m=mh$, the system (3.6.1) of $|\mathcal{R}(i)|$ linear integral equations is replaced by m systems of $|\mathcal{R}(i)|$ linear ordinary equations for the values of those functions at the time points $t_1, \ldots, t_m$. These systems of linear equations, whose coefficient matrices are in general sparse, can be solved by means of the Gauss–Seidel iteration method (for details we refer to Neumann and Steinhardt (1979a), section 2.3, and Schwarz (1981)). Since the functions Y_{ij} are increasing and

$$z_{ij} = \lim_{t \to \infty} Y_{ij}(t)$$

---

11     $G(a-)$ is the left–hand limit of G at the point a.

it suggests itself to choose m such that

(3.6.3) $\displaystyle\sum_{j\in Sn\mathcal{R}(i)} [z_{ij}-Y_{ij}(mh)] \leq \epsilon$ for all $i\in R$

where $\epsilon>0$ is a prescribed tolerance and z_{ij} is a solution to (3.6.2). For a STEOR
network it holds that

$$\sum_{j\in Sn\mathcal{R}(i)} z_{ij} = 1 \quad \text{for all } i\in R$$

Then (3.6.3) can be replaced by

(3.6.4) $\displaystyle 1 - \sum_{j\in Sn\mathcal{R}(i)} Y_{ij}(mh) \leq \epsilon$ for all $i\in R$

If the underlying network N is acyclic, it is again recommended to order the nodes of N
topologically so that node 1 is a source and $\mathcal{P}(j)\subseteq\{1,\ldots,j-1\}$ for each node $j>1$.
Hence, for fixed $i\in R$, the activation functions Y_{ij} can be determined successively for
increasing $j\in\mathcal{R}(i)$, that is, for the computation of Y_{ij} only functions Y_{ik} with $k<j$
(already found before) are needed.

Even if N contains cycles, the activation functions can be determined successively (this
approach is then called the **version II** of the MRP method) provided that the following
assumption is satisfied:

Assumption A0.

 There is a $\theta>0$ such that $F_{ij}(\theta)=0$ for every arc $<i,j>$ of the GERT network
 with $i\geq j$.

A0 says that the duration of the activities $<i,j>$ with $i\geq j$ is at least θ with probability
1. By an appropriate topological ordering of the nodes of the network, it can always be
guaranteed that there is no arc $<i,j>$ with $i\geq j$ outside cycles and that there are only
few arcs of that kind within cycles.

Owing to assumption A0 we have for $k \geq j$

$$F_{kj}(t) = 0 \text{ for } 0 \leq t \leq \theta \text{ or, respectively, } F_{kj}(t-s) = 0 \text{ for } t-\theta \leq s \leq t$$

and thus

$$\int_0^t F_{kj}(t-s)Y_{ik}(ds) = \begin{cases} \int_0^{t-\theta} F_{kj}(t-s)Y_{ik}(ds) & \text{for } t > \theta \\ 0 & \text{for } 0 \leq t \leq \theta \end{cases}$$

Substituting into (3.6.1) gives

(3.6.5) $$Y_{ij}(t) = p_{ij}F_{ij}(t) + \sum_{\substack{k \in \mathcal{R}(i) \cap \mathcal{P}(j) \\ k < j}} p_{kj} \int_0^t F_{kj}(t-s)Y_{ik}(ds)$$

$$+ \begin{cases} \sum_{\substack{k \in \mathcal{R}(i) \cap \mathcal{P}(j) \\ k \geq j}} p_{kj} \int_0^{t-\theta} F_{kj}(t-s)Y_{ik}(ds) & \text{for } t > \theta \\ 0 & \text{for } 0 \leq t \leq \theta \end{cases} \qquad (i \in R, j \in \mathcal{R}(i))$$

Then, for fixed $i \in R$, the following procedure proves to be expedient:

(3.6.6) Algorithm.
Step 1. Compute $Y_{ij}(t)$ for $0 \leq t \leq \theta$ by means of (3.6.5) successively for increasing $j \in \mathcal{R}(i)$

Step 2. Compute $Y_{ij}(t)$ for $\theta < t \leq 2\theta$ by means of (3.6.5) successively for increasing $j \in \mathcal{R}(i)$

Proceed in the same manner until, say after step m,

(3.6.7) $$\sum_{j \in S \cap \mathcal{R}(i)} [z_{ij} - Y_{ij}(m\theta)] \leq \epsilon$$

∎

In (3.6.7) $\epsilon > 0$ is again a prescribed tolerance and z_{ij} is a solution to (3.6.2). If the network in question is a STEOR network, (3.6.7) can be replaced by

$$1 - \sum_{j \in S \cap \mathcal{R}(i)} Y_{ij}(m\theta) \leq \epsilon$$

If assumption A0 is satisfied, the coefficient matrices of the systems of equations that have to be solved in version I of the MRP method reduce to lower–triangular matrices. Thus, only one iteration step of the Gauss–Seidel method is needed to solve such a system of equations. Numerical tests have shown that, if A0 holds, versions I and II approximately require the same computing time and storage capacity (for details we refer to Schwarz (1981)).

Chapter 4 Reducible GERT Networks

We have seen in chapter 3 that STEOR networks can be evaluated in a relatively easy manner by means of the MRP method. Nodes with AND entrance or IOR entrance, however, cause some difficulties. Therefore, it suggests itself to investigate GERT networks all of whose AND nodes and IOR nodes lie inside special subnetworks which can be reduced to structures containing only STEOR nodes. Obviously, the reduction of those subnetworks to "STEOR subnetworks" should be done such that the "reduced" GERT network is equivalent to the original ("reducible") GERT network.

In Neumann (1984c) it is also discussed how to reduce a subnetwork N' of a GERT network N to an "EOR subnetwork" (that is, nodes with deterministic exit and EOR entrance are permitted in addition to STEOR nodes). This reduction, however, is more complicated than the reduction to STEOR subnetworks and results in a GERT network which is equivalent to N only if some additional assumptions are satisfied. Moreover, to evaluate N "non–completely" (that is, to determine only the activation functions Y_{ij}, $j \in R, j \in S \cap \mathcal{Z}(i)$), the subnetwork N' generally has to be evaluated "completely" (that is, the functions $Y'_{i,\overline{S}(i)}, i \in R', \overline{S}(i) := \overline{S} \cap \mathcal{Z}(i), \overline{S} \in \mathbb{P}(S')$ must be computed, where $Y'_{i,\overline{S}}, R'$, and S' denote the activation functions, source set, and sink set, respectively, of N'). Therefore, this reduction is not discussed in what follows.

4.1 STEOR–Reducible Subnetworks

Recall the concept of a GERT subnetwork of a GERT network, which itself represents a GERT network (aside from its initial distribution) if it is treated as a separate entity (cf. section 1.4). In this section we deal with the evaluation of GERT subnetworks which behave like STEOR networks.

(4.1.1) **Definition.**
A GERT subnetwork N' of a GERT network N is called a **STEOR–reducible subnetwork** if the following condition is satisfied:

(a)   When a source i of N' is activated during an execution of N, then at most one and with probability 1 exactly one sink $j \in \mathcal{Z}(i)$ of N' is activated.

(4.1.2)   Remarks.

(a)   A STEOR–reducible subnetwork is not necessarily admissible (that is, it does not necessarily satisfy assumptions A4, A5 and A6 if treated as a separate entity). For example, in Fig. 2.2.2 network N_2 represents a nonadmissible STEOR–reducible subnetwork (if it is imbedded in some GERT network).

(b)   Condition (4.1.1a) corresponds to the basic property of a STEOR network (cf. theorem (3.2.11)). Thus, an admissible STEOR–reducible subnetwork taken on its own globally behaves like a STEOR network.

In analogy to (2.2.11), (3.5.11), and (3.5.12) let Y'_{ij} and z'_{ij} ($i \in R'$, $j \in S' \cap \mathcal{R}(i)$) be the activation function and activation number, respectively, of sink j given source i of an admissible STEOR–reducible subnetwork N' (where N' is considered a separate entity and R' and S' are the source set and sink set, respectively, of N'). By (3.5.16) we then have

(4.1.3)   Proposition.

An admissible STEOR–reducible subnetwork N' with source set R' and sink set S' is equivalent to the bipartite STEOR network with source set R', sink set S', and the arcs $\langle i,j \rangle$ ($j \in R'$, $j \in S' \cap \mathcal{R}(i)$) with weights $\begin{bmatrix} z'_{ij} \\ Y'_{ij}/z'_{ij} \end{bmatrix}$ [12].

A nonadmissible STEOR–reducible subnetwork may or may not be equivalent to a STEOR network. For example, in Fig. 2.2.2 the nonadmissible STEOR–reducible subnetwork N_2 is equivalent to the STEOR network N_1. In Fig. 4.1.1 the nonadmissible STEOR–reducible subnetwork N' (where each activity is again marked with its deterministic duration) is not equivalent to the STEOR network N'' (with appropriate duration of activity $\langle i',j' \rangle$) because node j in N' is activated twice whereas node j' in N'' is activated only once.

---

[12]   For $z'_{ij}=0$ we can choose any distribution function for the duration of activity $\langle i,j \rangle$, for example, the unit step function I_0.

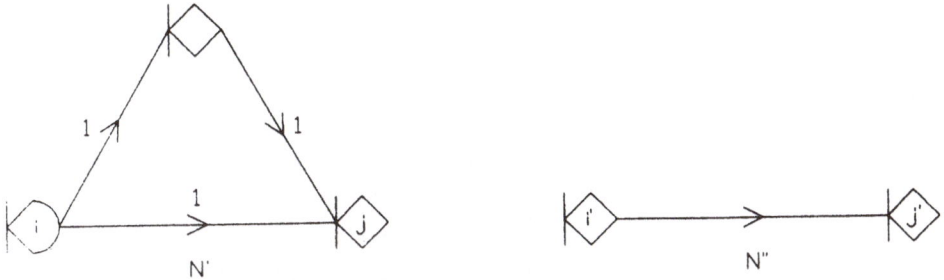

Figure 4.1.1

(4.1.4)   **Definition.**
An admissible STEOR–reducible subnetwork N' is said to be **evaluated** if the activation functions Y'_{ij} ($i \in R'$, $j \in S' \cap \mathcal{R}(i)$) have been computed.

(4.1.5) Let N be an admissible GERT network and N' be a STEOR–reducible subnetwork of N (note that N' is also admissible). If, after having evaluated N', N' is replaced by an equivalent bipartite STEOR network according to proposition (4.1.3), then the resulting GERT network $\bar{N}$ is again admissible and equivalent to N.

4.2 Cycle Reduction

As an applicaton of the evaluation of STEOR–reducible subnetworks and their replacement by simpler structures, we want to deal with the so–called cycle reduction in an admissible GERT network N which contains cycles. That is, we will replace N with an acyclic GERT network $\bar{N}$ equivalent to N. The methods for evaluating admissible GERT networks sketched in sections 2.4 and 4.6 contain such a cycle reduction as step 1.

(4.2.1) Let C be any cycle structure of an admissible GERT network N. We are going to expand C such that we obtain a complete subnetwork (cf. definition (1.4.3)). For each entrance node k of C, we add a STEOR node k' and a dummy activity <k',k> to C, where all entrance arcs of C with final node k now lead into node k'. Analogously, for each exit node l of C, we add a STEOR node l" and a dummy activity <l,l"> to C where all exit arcs of C with initial node l now emanate from node l" (observe that a node of C may be an entrance node and an exit node of C at the same time). The execution probability of activity <l,l"> equals the sum Σ of the execution probabilities of all exit arcs <1,s> of C with initial node 1, and the execution probability of an activity <1",s> equals the execution probability of arc <1,s> divided by Σ. The structure C' obtained from C by carrying out these modifications is called the **expanded cycle structure** belonging to C. The construction of such an expanded cycle structure is illustrated in Figs. 4.2.1 and 4.2.2 (where the arc weights are omitted). If we replace C with C' in the underlying GERT network N, we obtain an admissible GERT network N̂, which is equivalent to N.

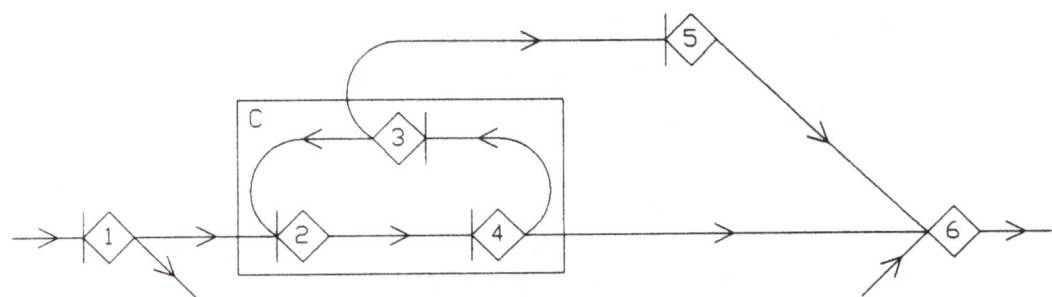

Figure 4.2.1

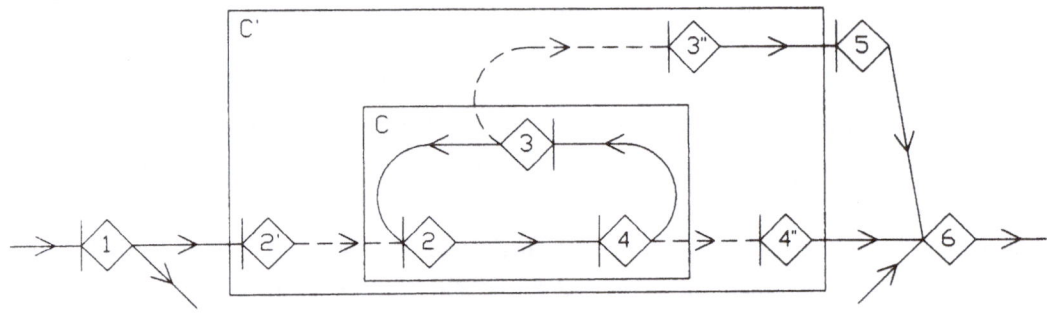

Figure 4.2.2

The expanded cycle structure C' satisfies conventions (1.4.1) and (1.4.2) and thus represents a complete subnetwork (cf. definition (1.4.3)). The source set R' of C' consists of the initial nodes k' of the dummy activities <k',k> added to C, and the sink set S' consists of the final nodes 1" of the dummy activities <1,1">. C' also satisfies condition (1.4.4a) in the definition of a GERT subnetwork because the underlying GERT network fulfills assumption A6. Moreover, condition (4.1.1a) is met because, owing to assumption A4, C and C' contain only STEOR nodes. Hence, C' is a STEOR–reducible subnetwork of N̂.

(4.2.2) The activation functions Y'_{ij} and activation numbers z'_{ij} in the expanded cycle structure C' ($i \in R'$, $j \in S'$) can be determined by the MRP method. Then C' can be replaced by the complete bipartite STEOR network C'' with source set R' and sink set S', where arc $\langle i,j \rangle$ ($i \in R'$, $j \in S'$) has weight $\begin{bmatrix} z'_{ij} \\ Y'_{ij}/z'_{ij} \end{bmatrix}$. Fig. 4.2.3 illustrates that replacement for the expanded cycle structure of Fig. 4.2.2, where we have again omitted the arc weights.

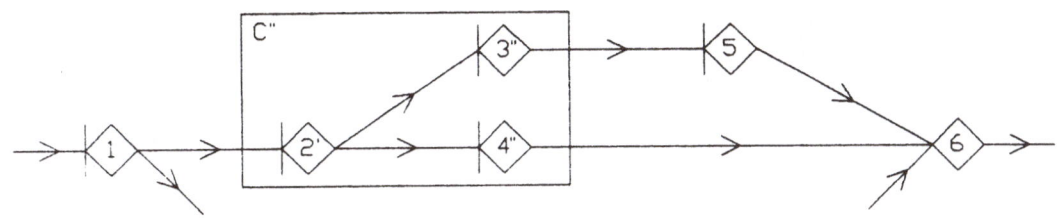

Figure 4.2.3

If we expand each cycle structure of the underlying admissible GERT network N in the aforementioned way and replace it as just explained, we obtain an acyclic GERT network N̄, which is again admissible and equivalent to N.

4.3 Nodes Which Belong Together

In what follows, the GERT network N in question is always supposed to satisfy assumptions A4 and A5.

(4.3.1) Owing to convention (1.3.11), for each (genuine) IOR node j of N there is a network realization in which at least two of the arcs leading into j are carried out. Hence, due to assumption A1, for each IOR node j there is a deterministic node i in N from which *at least two* distinct predecessors of j are reachable. That is, $|\mathcal{P}_i(j)| \geq 2$ where $\mathcal{P}_i(i) := \mathcal{P}(j) \cap \mathcal{R}(i)$ is the set of those predecessors of j which are reachable from i. In this case we say that node j fits node i. Note that there may be predecessors of node j which are not reachable from i, and there may be walks emanating from i which do not lead into j (see Fig. 4.3.1).

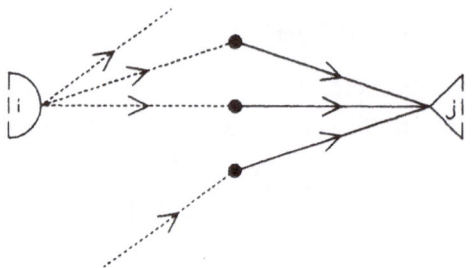

Figure 4.3.1 [13]

(4.3.2) Now let j be an AND node of N. In order that node j can be activated at all, there has to be a deterministic node i in N from which *all* predecessors of j are reachable, that is, $P_i(j)=P(j)$. In this case we also say that node j **fits** node i (compare Fig. 4.3.2).

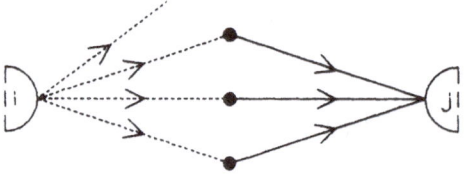

Figure 4.3.2

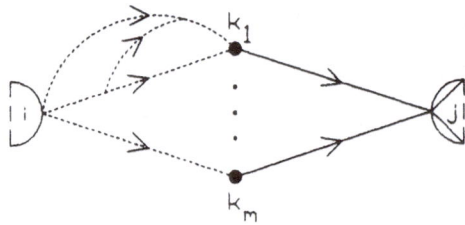

Figure 4.3.3 [14]

Let i be a deterministic source and $j\in Z(i)$ be an AND sink or IOR sink of an admissible STEOR–reducible subnetwork N'. Later we will see that the activation function Y'_{ij} in N' is relatively easy to compute if all walks from i to distinct

---

13   Dotted–line arrows again represent walks.

14   The symbol for node j means a node with AND entrance or IOR entrance.

predecessors of j are disjoint aside from their common initial node (see Fig. 4.3.3). This leads to the following

(4.3.3) Definition.

Let i be a deterministic node and $j\in\ell(i)$ be an AND node or IOR node of a GERT network. Then j is said to **belong to** i (and i and j are called nodes **belonging together**) if

(a)   j fits i

(b)   Walks from i to distinct predecessors of j are disjoint aside from their common initial node.

(4.3.4) The set of the nodes belonging to a deterministic node i is designated by $B(i)$. If j is an AND node or IOR node, then $\bar{B}(j)$ is the set of those nodes to which j belongs. For the GERT network (without arc weights) in Fig. 4.3.4 we have

$$B(1) = \{3,5\}, \ B(2) = \{4\}$$
$$\bar{B}(3) = \bar{B}(5) = \{1\}, \ \bar{B}(4) = \{2\}$$

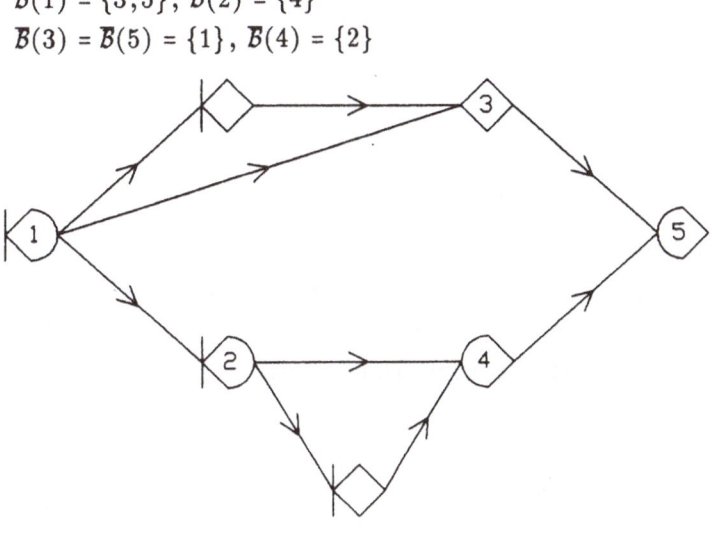

Figure 4.3.4

Now let i be a deterministic source and $j\in B(i)$ be a sink of an admissible STEOR–reducible subnetwork N'. Furthermore, let $k_1,\dots,k_m$ be the predecessors of j which are reachable from i (that is, $P_i(j)=\{k_1,\dots,k_m\}$, compare Fig. 4.3.3), let T_i and T_j be the times of activation of nodes i and j, respectively, and let Z_μ be the time of termination of activity $<k_\mu, j>$ $(\mu=1, \ldots, m)$. Taking assumption A2b into account, which says roughly speaking that the passing times of disjoint walks with common initial node are independent of each other, and using the fact that for m independent random events $A_1, \ldots, A_m$ it holds that

$$P(\bigcap_{\mu=1}^{m} A_\mu) = \prod_{\mu=1}^{m} P(A_\mu)$$

$$P(\bigcup_{\mu=1}^{m} A_\mu) = 1 - \prod_{\mu=1}^{m} [1-P(A_\mu)]$$

we obtain for $t \in \mathbb{R}$

$$(4.3.5) \quad Y'_{ij}(t) = P(T_j - T_i \le t \mid A_i) = \begin{cases} \prod_{\mu=1}^{m} P(Z_\mu - T_i \le t \mid A_i) & \text{for an AND node } j \\ 1 - \prod_{\mu=1}^{m} [1 - P(Z_\mu - T_i \le t \mid A_i)] & \text{for an IOR node } j \end{cases}$$

To compute the activation function Y'_{ij} in N' it remains to be determined the probabilities $P(Z_\mu - T_i \le t \mid A_i)$ for $\mu=1, \ldots, m$, which can be done with relative ease in the case that N' represents a so-called basic element structure. Basic element structures will be discussed in sections 4.4 to 4.6. The theory and evaluation of GERT networks with basic element structures is treated in detail in Neumann and Steinhardt (1979a). Thus we will only sketch the most important results.

4.4 Basic Element Structures

In what follows we again stipulate that the GERT network N in question always satisfies assumptions A4 and A5 (if, in addition, the "hard" assumption A6 is required, we will mention this explicitly). A basic element structure represents a complete subnetwork N' of N (cf. definition (1.4.3)) which, among other things, has the following properties:

(i) All nodes of N belonging to a deterministic source of N' are sinks of N' and, vice versa, all nodes of N to which an AND sink or IOR sink of N' belongs are sources of N'.

(ii) Each AND sink or IOR sink of N' reachable from a deterministic source i of N' belongs to i.

To define a basic element structure precisely let R' and S' be again the source set and sink set, respectively, of N', Moreover, let R'_D and R'_{St} be the set of the deterministic sources and stochastic sources, respectively, of N', and let S'_A be the set of the sinks of N' with AND entrance and S'_{AI} be the set of the sinks of N' with AND entrance or IOR entrance. Then the definition of a basic element structure is as follows.

(4.4.1) **Definition.**

A weakly connected complete subnetwork N' of a GERT network (that fulfills A4 and A5) is called a **basic element structure** if the following conditions are satisfied:

(a) At least one source of N' is deterministic

(b) For all $i \in R'_D$ we have $\emptyset \neq B(i) \subseteq S'_{AI}$

   For all $j \in S'_{AI}$ we have $\emptyset \neq B(j) \subseteq R'_D$

(c) If $i \in R'_D, j \in S'_{AI}$, and $j \in \mathcal{R}(i)$, then $j \in B(i)$

(d) For each $i \in R'_{St}$ we have $\mathcal{R}(i) \cap S'_A = \emptyset$

(e) If $i_1, i_2 \in R'$ where $i_1 \neq i_2$, then $i_2 \notin \mathcal{R}(i_1)$ and $i_1 \notin \mathcal{R}(i_2)$

   If $j_1, j_2 \in S'$ where $j_1 \neq j_2$, then $j_2 \notin \mathcal{R}(j_1)$ and $j_1 \notin \mathcal{R}(j_2)$

(f) If a source i of N' is activated, then at most one and with probability 1 exactly one sink $j \in \mathcal{R}(i)$ of N' is activated.

(4.4.2) **Remarks.**

(a) From (4.4.1a) and (4.4.1b) it follows that each basic element structure contains at least one sink with AND entrance or IOR entrance.

(b) Conditions (4.4.1b) and (4.4.1c) correspond to the properties (i) and (ii) listed above.

(c) Conditions (4.4.1c) and (4.4.1d) guarantee that the "transition probability" from any source i to any sink $j \in \mathcal{R}(i)$ is positive if the execution probabilities of all activities belonging to walks from i to j are positive and if these walks contain only EOR nodes.

(d)   By definition no source and no sink of a basic element structure N' belong to a cycle that lies entirely in N'. Condition (4.4.1e) excludes the case that there are two distinct sources or two distinct sinks of N' which belong to a cycle C where C does not lie entirely in N' (see Fig. 4.4.1). Observing the weak connectedness of N', assumption A4, and the fact that N' contains at least one source $i \in R'_D$ and at least one sink $j \in S'_{AI}$, it also follows from condition (4.4.1e) that there is no cycle that contains only one of the sources and only one of the sinks of N' (see Fig. 4.4.2). Thus, all sources and sinks of N' lie outside of cycles, in particular, each walk from a source of N' to a sink of N' belongs entirely to N'.

(e)   Condition (4.4.1f) coincides with condition (a) in definition (4.1.1).

Figure 4.4.1   Figure 4.4.2

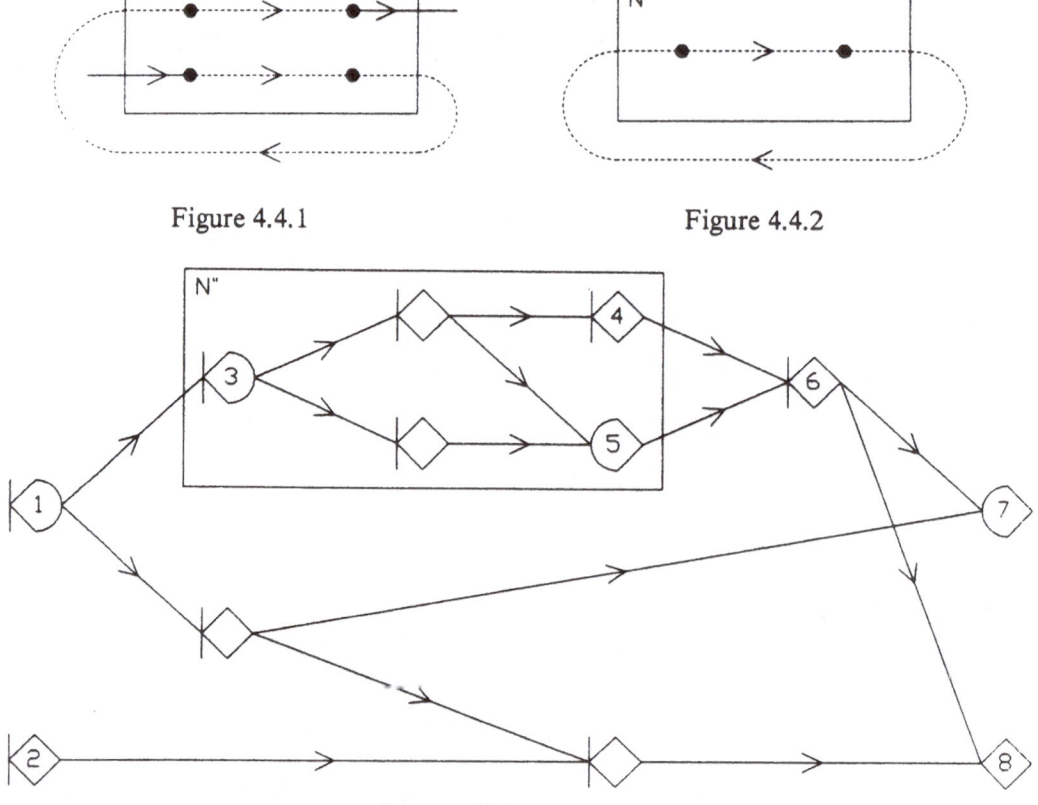

Figure 4.4.3. Basic element structure N'

Fig. 4.4.3 shows a basic element structure N' (again without arc weights). Inside the frame there is another basic element structure N" nested into N'. We have

$$B(1) = \{7,8\}, \; \overline{B}(7) = \overline{B}(8) = \{1\}$$
$$B(3) = \{5\}, \; \overline{B}(5) = \{3\}$$

The basic element structure N' has the sources 1 and 2 and the sinks 7 and 8. Basic element structure N" has the source 3 and the sinks 4 and 5. If AND node 5 is replaced by an IOR node, then N" does not represent a basic element structure because both sinks 4 and 5 may be activated during one and the same network realization (in addition, EOR node 6 has to be replaced by an IOR node in the latter case).

Next, we stipulate that the following convention is satisfied.

(4.4.3)   Convention.

(a)   Each stochastic source of a basic element structure N' has exactly one successor, and this successor is reachable from at least one deterministic source of N'.

(b)   Each EOR sink of a basic element structure N' has exactly one predecessor, and at least one AND sink or IOR sink of N' is reachable from this predecessor.

The basic element structures N' and N" in Fig. 4.4.3 fulfill convention (4.4.3). It is easy to see that each admissible GERT network can be transformed into an equivalent admissible GERT network all of whose basic element structures satisfy convention (4.4.3). This transformation is automatically carried out within the evaluation method sketched in section 4.6. An example is illustrated in Fig. 4.4.4 where N represents an admissible GERT network and a basic element structure that does not satisfy convention (4.4.3a). N is equivalent to the admissible GERT network $\hat{N}$, and the basic element structure N' of $\hat{N}$ meets condition (4.4.3a).

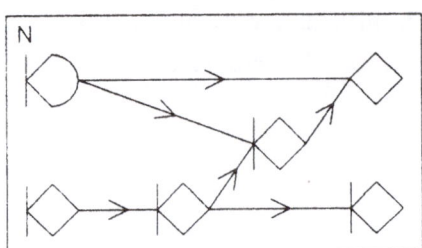 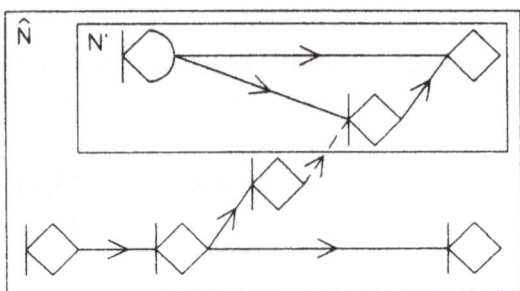

Figure 4.4.4

Using convention (4.4.3), the following theorem can be proved (cf. Neumann and Steinhardt (1979a), section 3.3.2):

(4.4.4)   Theorem.
If two basic element structures N' and N'' of a GERT network possess one and the same deterministic source or one and the same sink with AND entrance or IOR entrance, then N' and N'' coincide.

Theorem (4.4.4) turns out to be useful for finding the basic element structures in a given GERT network, which needs to be done within the evaluation method of section 4.6.

The following assumption, which corresponds to condition (1.4.4a), ensures that each basic element structure of a GERT network N is a GERT subnetwork (cf. definition (1.4.4)) and, owing to condition (4.4.1f), also a STEOR–reducible subnetwork of N.

Assumption A7.
During each execution of a GERT network N, at most one of those sources of each basic element structure N' of N is activated from which one and the same sink of N' is reachable.

(4.4.5)   Proposition.
Let N be a GERT network that satisfies assumptions A4, A5 and A7. Then each basic element structure of N represents a STEOR–reducible subnetwork of N.

Since a STEOR–reducible subnetwork is not necessarily admissible (cf. remark (4.1.2a)), a basic element structure N' of a GERT network N that satisfies assumption A7 need not be admissible, either (that is, N' does not necessarily represent an admissible GERT network if taken on its own) except that N satisfies assumption A6 (in addition to A4 and A5) and is thus admissible itself. Next we show how to test the admissibility of a basic element structure. This will prove to be useful later when we want to circumvent the hard testing of assumption A6 for certain GERT networks. We formulate a condition which is related to condition (3.3.1) for EOR networks.

(4.4.6)   Condition.

Let i be a deterministic node of a basic element structure N' with node set V' and let k_1 and k_2 be any two distinct successors of i. Then

(a) $\mathcal{R}(k_1) \cap \mathcal{R}(k_2) \cap V' \subseteq S'_{AI}$ if i is a source of N'

(b) $\mathcal{R}(k_1) \cap \mathcal{R}(k_2) \cap V' = \emptyset$ if i is not a source of N'.

Condition (4.4.6) says, figuratively speaking, that different walks separating at a deterministic node i of a basic element structure N' do not meet within N' except at a node from S'_{AI}, and the latter happens only if i is a source of N'. It can be tested in $O(|V|^2)$ time whether condition (4.4.6) is satisfied for each basic element structure of a given GERT network. The same reasoning as in the proof of theorem (3.3.2) gives

(4.4.7)   Theorem.

Let N be a GERT network that satisfies assumptions A4, A5 and A7. Then each basic element structure of N which meets condition (4.4.6) is admissible.

For a detailed proof we refer to Neumann and Steinhardt (1979a), section 3.3.2. Note that condition (4.4.6) is not necessary for the admissibility of a basic element structure. For example, node 3 in Fig. 4.4.3, which is not a source of the admissible basic element structure N', does not satisfy condition (4.4.6b).

4.5 BES Networks

The following assumption together with A7 ensures that admissible GERT networks all of whose AND nodes and IOR nodes belong to basic element structures can be reduced to admissible EOR networks.

Assumption A8.

> Every AND node and every IOR node of a GERT network N is a sink of a
> basic element structure of N.

Next, we define

(4.5.1)   Definition.

> An admissible GERT network that satisfies assumptions A7 and A8 is called a
> **BES network.**

Since each basic element structure of an admissible GERT network N that satisfies
assumption A7 is an admissible STEOR–reducible subnetwork of N, we obtain by
proposition (4.1.3) and (4.1.5)

(4.5.2)   Theorem.

> A BES network is equivalent to an admissible EOR network.

Using theorems (3.3.2) and (4.4.7) we get

(4.5.3)   Proposition.

Let N be a GERT network that satisfies assumptions A4, A5, A7, and A8. Assume that
condition (4.4.6) is met inside each basic element structure of N and condition (3.3.1) is
met outside every basic element structure of N. Then N also satisfies assumption A6 and
thus represents a BES network.

Proposition (4.5.3) enables us to circumvent the hard test of assumption A6. The
following theorem, which is an addition to theorem (4.4.4), is important to the
evaluation of BES networks. For the proof, which exploits, among other things,
condition (4.4.1f) and convention (4.4.3), we refer to Neumann and Steinhardt (1979a),
section 3.3.2.

(4.5.4)   Theorem.

> For two basic element structures of a BES network, either one is a subnetwork
> of the other or they are arc–disjoint (that is, they do not have any arcs in
> common).

Note that theorem (4.5.4) permits a STEOR node i to belong to two different basic element structures where i is a sink of one of the basic element structures and, at the same time, a source of the other basic element structure.

(4.5.5)   **Definition.**

A basic element structure N' of a GERT network N is called an **interior basic element structure** if

(a)   N' contains only EOR nodes except for its sinks

(b)   For every deterministic node i of N' which is not a source of N' it holds that $B(i)=\emptyset$ in N.

The basic element structure N" inside the frame in Fig. 4.4.3 is an interior basic element structure. Fig. 4.5.1 shows a GERT network N, which also represents an interior basic element structure and where node i is a deterministic node "in the interior" of N.

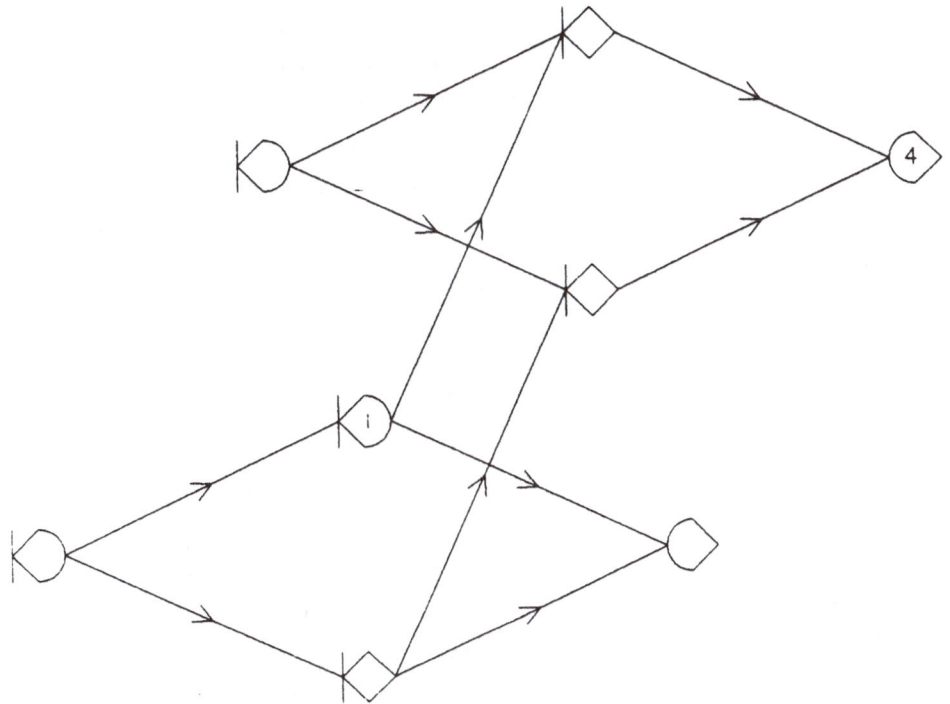

Figure 4.5.1

(4.5.6)   Remarks.

(a)   An interior basic element structure N' of a GERT network N does not contain any other basic element structure of N (that is, there exists no basic element structure different from N' which is a subnetwork of N'). On the other hand, a basic element structure of a BES network N that does not contain any other basic element structure is an interior basic element structure. Note that the latter need not hold if N does not represent a BES network as it is shown by the admissible GERT network in Fig. 4.5.2 (which violates assumption A8). The network N of Fig. 4.5.2 is a basic element structure with source 1 and sink 6, but neither does it contain any other basic element structure (the subnetworks N' or N'' inside the frames do not represent basic element structures because $B(2)=\{3,4,5\}$ in N but node 5 is no sink of N'' and nodes 3 and 4 are no sinks of N') nor is it an interior basic element structure (node 2 violates condition (4.5.5b) and nodes 3, 4 and 5 violate (4.5.5a)).

(b)   By theorem (4.5.4) each BES network that contains at least one AND or IOR node possesses at least one interior basic element structure.

(c)   To evaluate a BES network N we may proceed as follows: We evaluate the basic element structures of N and replace them by bipartite STEOR subnetworks successively from the inside outwards (note that by theorem (4.5.4) there are no "overlapping" basic element structures which have at least one arc or one non–STEOR node in common). The resulting network represents an admissible EOR network, which can be evaluated by means of the MRP method.

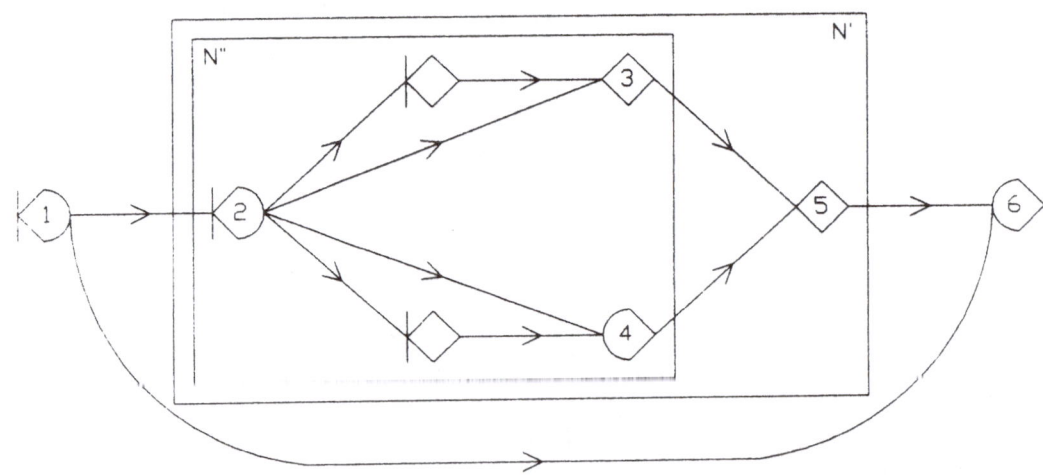

Figure 4.5.2

4.6 Evaluation Methods for BES Networks and General Admissible GERT Networks

As stated in remark (4.5.6c), basic element structures are to be evaluated successively from the inside outwards beginning with an interior basic element structure. In this section we first show how to evaluate an admissible interior basic element structure N' which represents a STEOR–reducible subnetwork, that is, how to compute the activation functions Y'_{ij} of N' for $i \in R'$, $j \in S' \cap \mathcal{R}(i)$ (for details we refer to Neumann and Steinhardt (1979a), section 3.4). Recall that for a BES network N, each basic element structure of N automatically represents an admissible STEOR–reducible subnetwork. Second, we sketch two methods for evaluating BES networks and general admissible GERT networks.

Let i be any source of an admissible interior basic element structure N'. To compute the activation functions Y'_{ij} for $j \in S' \cap \mathcal{R}(i)$, we consider the GERT subnetwork $N'(i)$ of N' induced by the node set $V' \cap \mathcal{R}(i)$, cf. (1.4.5). Since N' is an interior basic element structure, each node of $N'(i)$ with AND entrance of IOR entrance is a sink of $N'(i)$. Observing condition (4.4.1d) it can be shown that if the source i of $N'(i)$ is stochastic, then all sinks of $N'(i)$ are non–genuine IOR nodes (when $N'(i)$ is considered a separate entity) which have to be replaced by EOR nodes (cf. theorem (3.3.14) in Neumann and Steinhardt (1979a)). If i is a deterministic node, the sinks of $N'(i)$ may have AND, IOR or EOR entrance. Moreover, by remark (4.4.2d) each walk from source i to any sink of $N'(i)$ belongs entirely to N'.

The activation functions Y'_{ik} of all EOR nodes k of $N'(i)$ can be determined by the MRP method. The activation function Y'_{ij} for an AND sink or IOR sink j of $N'(i)$ is computed as follows: Let $k_1, \ldots, k_m$ be those predecessors of j in N' which are reachable from node i and thus belong to $N'(i)$ [15], and let Z_μ be the time of termination of activity $\langle k_\mu, j \rangle$ ($\mu = 1, \ldots, m$). Since the time elapsed between the activation of node i and termination of activity $\langle k_\mu, j \rangle$ is equal to the sum of the time between the activation of nodes i and k_μ and the time between the activation of node

_____

[15] If j is an IOR node, then not necessarily all predecessors of j in N' belong to $N'(i)$.

k_μ and termination of activity $<k_\mu, j>$, we obtain

$$P(Z_\mu - T_i \leq t \mid A_i) = P_{k_\mu j} \int_{[0,t]} F_{k_\mu j}(t-s) Y'_{ik_\mu}(ds) \quad (\mu=1,\ldots,m; t \geq 0)$$

Relation (4.3.5) eventually provides $Y'_{ij}(t)$.

After the computation of the activation functions Y'_{ij} $(i \in R', j \in S' \cap \mathcal{R}(i))$, the basic element structure N' can be replaced by the bipartite STEOR network with source set R', sink set S', and arcs $<i,j>$ $(i \in R', j \in S' \cap \mathcal{R}(i))$ with weights $\begin{bmatrix} z'_{ij} \\ Y'_{ij}/z'_{ij} \end{bmatrix}$, cf. proposition (4.1.3). In this manner we can reduce the interior basic element structures of a BES network successively to STEOR subnetworks. As already stated in remark (4.5.6c), basic element structures one nested into another are evaluated and replaced "from the inside outwards". By this procedure we finally obtain an admissible EOR network, which can be evaluated by the MRP method.

The successive reduction of basic element structures is illustrated in Fig. 4.6.1 where the BES network N contains the three basic element structures N_1, N_2 and N_3. At first the interior basic element structures N_1 and N_2 are reduced resulting in the network $\bar{N}$ with the interior basic element structure $\bar{N}_3$. The reduction of $\bar{N}_3$ gives the final network $\bar{\bar{N}}$.

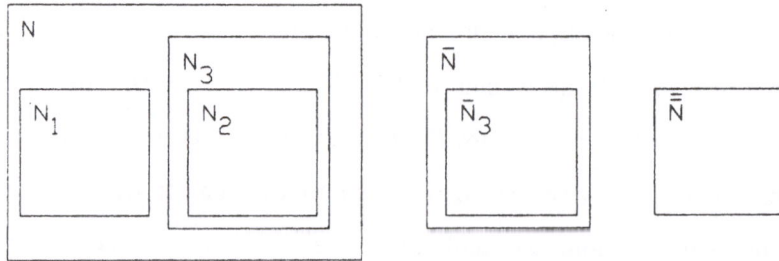

Figure 4.6.1

In Neumann and Steinhardt (1979a), section 3.5, an algorithm is presented which is based upon the aforementioned considerations. This procedure is called the **BES method** because it evaluates each BES network. In addition, certain admissible GERT networks that do not represent BES networks can be evaluated with the aid of this method. The BES method requires that the GERT network under consideration satisfies the assumptions A1 through A5, which can be checked relatively easily aside from A2 (which stipulates the independence of the random variables describing the stochastic behaviour of the GERT network and which is supposed to be fulfilled a priori). Whether the remaining assumptions A6, A7 and A8 hold, is examined in the course of the procedure.

In what follows, we briefly sketch the BES method. The procedure consists of two parts. In part 1 it is tested whether the BES method can really evaluate the GERT network N in question. To this end, for each deterministic node i of N, the set $B(i)$ of the nodes belonging to i is determined by means of a labelling process. With the aid of the sets $B(i)$ the basic element structures of N can be found successively. In addition, some subnetworks that do not represent basic element structures (for example, the subnetworks that satisfy conditions (4.4.1a) to (4.4.1f) but not convention (4.4.3)) are transformed into basic element structures by introducing dummy activities and auxiliary nodes. This provides a GERT network equivalent to N. Simultaneously, the assumptions A6, A7 and A8 are examined [16] and the non–genuine IOR nodes are determined and replaced by EOR nodes. Since part 1 of the BES method only comprises some structural analysis of the underlying GERT network, the distribution functions of the activity durations are not needed in that part.

In case it turns out that the GERT network N in question cannot be reduced to an admissible EOR network, the procedure is discontinued. Otherwise in part 2 the basic element structures of N are evaluated and replaced by bipartite STEOR subnetworks successively and the MRP method is applied to the resulting network. Numerical tests have shown that, on the average, about 10 per cent of the total computation time are needed for part 1 of the BES method so that a premature termination of the procedure is no great loss (cf. Heine (1979)).

---

[16]   For example, A7 and A8 are not satisfied if we discover some contradiction to theorem (4.5.4), and the testing of assumption A6 can be circumvented owing to proposition (4.5.3).

Simulation techniques [17] generally give worse results than the BES method, in particular, if the GERT network in question contains a large number of cycles and there are activities within cycles whose execution probabilities are very small or near 1. In the former case we need a large number of simulation runs to obtain results which are not too inaccurate. In the latter case one generally cycles many times within the network until the realizable sinks are reached during a simulation run, which also causes a considerable increase in computing time.

In practice, GERT networks mostly contain several basic element structures and STEOR subnetworks, which can be evaluated very efficiently using the BES and MRP methods. Thus, it suggests itself to combine simulation with those efficient methods for the evaluation of general GERT networks. Such a **universal method** is presented in Neumann and Steinhardt (1979a), section 7.5. This method can be applied to every admissible GERT network, where assumption A6 is examined and non–genuine IOR nodes are discovered in the course of the procedure. The algorithm is as follows:

(4.6.1)   Algorithm.

Step 1.   The cycle structures of the underlying GERT network N are determined and reduced to acyclic STEOR subnetworks as described in section 4.2. This gives an acyclic GERT network $\bar{N}$.

Step 2.   The basic element structures of $\bar{N}$ are determined and, in case they represent STEOR–reducible subnetworks, evaluated and successively replaced by bipartite STEOR subnetworks.

Step 3.   STEOR subnetworks are determined, evaluated and replaced by ("smaller") bipartite STEOR subnetworks. This provides a further reduction in size of the GERT network.

Step 4.   If the resulting GERT network represents an EOR network, it can be evaluated by the MRP method; otherwise it is evaluated by means of simulation.

■

---

[17]   For simulation techniques we refer to Neumann and Steinhardt (1979a), chapter 7, Pritsker (1977), Pritsker and Sigal (1983), and Whitehouse (1973).

Note that the GERT network obtained at the end of step 3 is acyclic and in general much "smaller" then the original network. Numerical tests have shown that the universal method mostly requires less computation time than a simulation technique applied to the original GERT network (cf. Fix (1979)).

Chapter 5   Scheduling with GERT Precedence Constraints

Scheduling can be loosely defined as the art of assigning resources to tasks in order to insure the termination of the tasks in a reasonable amount of time or to minimize a certain cost function. Since the terminology of scheduling theory arose in the processing and manufacturing industries, one generally speaks of "machines" instead of resources and of "jobs" instead of tasks that have to be "processed" by the machines.

In this chapter, we will first consider some simple deterministic single–machine scheduling problems and introduce some basic concepts (for further information we refer to Baker (1974) or Lawler et al. (1982)). Then we will deal with single–machine scheduling problems where the precedence constraints for the jobs are given by a GERT network. By exploiting special properties of admissible EOR networks, polynomial algorithms for two scheduling problems with EOR precedence constraints will be derived. Min–sum scheduling problems with general GERT precedence networks can be reduced to problems of stochastic dynamic programming. Most of the following material can be found in Bücker (1990), Bücker and Neumann (1989), Bücker et al. (1990), and Rubach (1984).

As we will see, all single–machine scheduling problems with GERT precedence constraints are NP–hard aside from few special cases. Thus there is no hope of finding efficient exact algorithms for solving multi–machine scheduling problems. For the latter problems, only the method of simulation and heuristics seem to be appropriate which are beyond the scope of this monograph.

5.1   Deterministic Single–Machine Scheduling

We consider the following simple sequencing problem. There are n jobs numbered $1,\ldots,n$ to be processed by a single machine which can execute at most one job at a time. Each job ν requires a (nonnegative) **processing time**, also called **duration** of the job and denoted by D_ν $(\nu=1,\ldots,n)$. The jobs are supposed to be executed without interruption and without idle times between them with the first job beginning at time 0. Moreover, each job is assumed to be available for processing at time 0. Then any given sequence of jobs induces a well–defined **completion time** C_ν for each job ν $(\nu=1,\ldots,n)$. Fig. 5.1.1 shows four jobs processed without interruption and idle times.

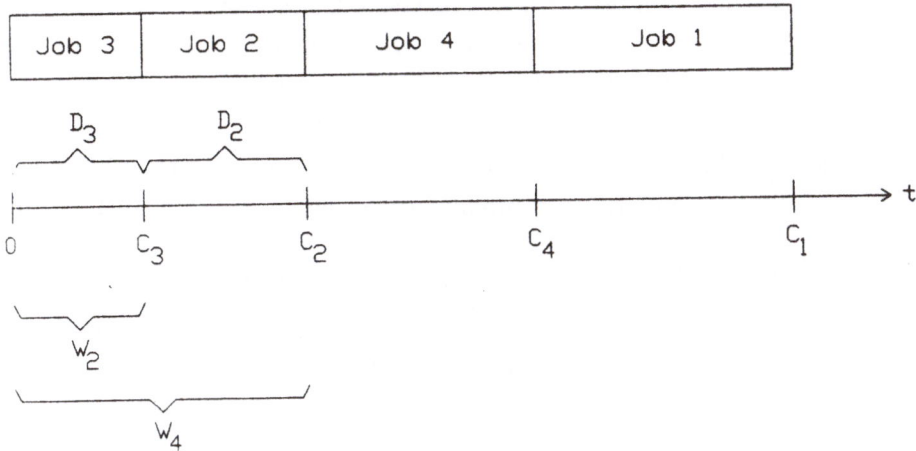

Figure 5.1.1

Suppose that the objective is to find a sequence of jobs, also called a **list schedule**, that minimizes the sum of job completion times $\sum_{\nu=1}^{n} C_\nu$. This is equivalent to minimizing the mean **flow–time** $\frac{1}{n} \sum_{\nu=1}^{n} C_\nu$, that is, the mean amount of time a job spends in the system. Let W_ν be the waiting time of job ν preceding its operation (see Fig. 5.1.1). Then

$$\sum_{\nu=1}^{n} C_\nu = \sum_{\nu=1}^{n} (D_\nu + W_\nu)$$

Since $\sum_{\nu=1}^{n} D_\nu$ is a constant for all job sequences, we have to minimize $\sum_{k=1}^{n} W_\nu$. To find a list schedule $(\nu_1, \nu_2, \ldots, \nu_n)$ that minimizes $\sum_{k=1}^{n} W_{\nu_k}$, where the sequence $(W_{\nu_1}, \ldots, W_{\nu_n})$ is nondecreasing, we have to make each W_{ν_k} as small as possible. It holds that $W_{\nu_1} = 0$ and $W_{\nu_2} = D_{\nu_1}$ because job ν_1 starts at time 0 and job ν_2 must wait only for ν_1 to be processed. Hence, if we choose ν_1 to have the shortest processing time of all the jobs $\nu_1, \ldots, \nu_n$, we minimize W_{ν_2}. Since $W_{\nu_3} = D_{\nu_1} + D_{\nu_2}$, we choose ν_1 and ν_2 to have the shortest and next shortest processing time from the jobs $\nu_1, \ldots, \nu_n$. In other words, job ν_2 has the shortest processing time of the "remaining" jobs $\nu_2, \ldots, \nu_n$. Continuing in this way we obtain the result that any list schedule is optimal which places the jobs in nondecreasing order of processing times. This is the so–called **SPT rule**, where "SPT" means that the "shortest processing time" comes first. The list schedule in Fig. 5.1.1 satisfies the SPT rule and is thus optimal.

(5.1.1) Now let $w_\nu \geq 0$ be a weighting factor describing the urgency or importance of job ν ($\nu=1,\ldots,n$). Suppose we wish to minimize the weighted sum of completion times or respectively mean **weighted flow–time** $\sum\limits_{\nu=1}^{n} w_\nu C_\nu$. A similar reasoning as above shows that this problem is solved by the following **ratio rule** of Smith (cf. Smith (1956)): Any list schedule is optimal that puts the jobs in order of nondecreasing ratios

$$\rho_\nu := \frac{D_\nu}{w_\nu} \quad (1 \leq \nu \leq n)$$

(5.1.2) Next, we assume that **precedence constraints for the jobs** are given, where "job k precedes job l" means that job k has to be completed before job l can be begun. Such a precedence relation can be represented by an acyclic directed graph G where the jobs correspond to the nodes and job k precedes job l exactly if $l \in \ell(k)$ in G. We consider the scheduling problem of minimizing the weighted sum of job completion times, where the precedence constraints for the jobs are given by an outtree (compare (1.1.3)). This scheduling problem is designated by the symbol $1|\text{tree}|\Sigma w_\nu C_\nu$, where "1" is the number of machines, "tree" describes the precedence constraints, and the third field contains the objective function. Exploiting the fact that each node in an outtree different from the root has exactly one predecessor, Smith's rule can be generalized to solve problem $1|\text{tree}|\Sigma w_\nu D_\nu$ as follows (cf. Adolphson and Hu (1973), and Horn (1972)):

Let $\{1,\ldots,n\}$ be again the set of all jobs (identified with the nodes of the outtree) and let J be the set of the jobs different from the root of the outtree. Then we have the following algorithm:

(5.1.3) Find a job $l \in J$ such that

$$\rho_l = \min_{\nu \in J} \rho_\nu$$

Let k be the (immediate) predecessor of 1. Then the subsequence $(k,1)$ appears in at least one optimal list schedule and can be treated as one **composite job** with processing time D_k+D_1, weight w_k+w_1, and set of successors $\mathcal{S}(k)\cup\mathcal{S}(1)\setminus\{1\}$. Suppose that the composite job is again denoted by k. Delete 1 from J and go to (5.1.3) if $J\neq\emptyset$ (otherwise terminate).

The algorithm can be implemented to run in $0(n \log n)$ time. Later we will see that owing to the tree–structure property of admissible EOR networks, the algorithm can be generalized to the case of a scheduling problem with precedence constraints given by an admissible EOR network.

(5.1.4) Finally, we consider the single–machine scheduling problem which consists of finding a list schedule that minimizes the **maximum job completion cost**, where the precedence constraints for the jobs are given by an arbitrary acyclic directed graph. Let f_ν be a nondecreasing real–valued function where $f_\nu(t)$ represents the cost arising when job ν is completed at time t. The maximum job completion cost is then

$$\max_{\nu=1, \ldots, n} f_\nu(C_\nu) =: f_{max}$$

This scheduling problem is denoted by the symbol $1|\text{prec}|f_{max}$ and can be solved by successively applying **Lawler's rule** which is as follows (cf. Lawler (1973)): From among all unscheduled jobs without successors, put that job last (that is, insert it at the top of the current list schedule) which will incur the smallest cost in that position. In detail, Lawler's rule says the following: Let $J\subseteq\{1,\ldots,n\}$ be the set of the unscheduled jobs, let $S\subseteq J$ be the set of the jobs from J without successors in J, and let $D_J:= \sum_{\nu\in J} D_\nu$. Then job $1\in S$ with

$$f_1(D_J) = \min_{\nu\in S} f_\nu(D_J)$$

is to be processed after the execution of all jobs from $J\setminus\{1\}$ and is deleted from J. By successive application of this procedure an optimal list schedule is constructed backwards. If each function value $f_\nu(t)$ is evaluated in unit time, the time complexity of Lawler's algorithm is $0(n^2)$.

(5.1.5) A special case of problem $1|\text{prec}|f_{\text{max}}$ is the problem $1|\text{prec}|L_{\text{max}}$ where

$$f_{\nu}(C_{\nu}) = L_{\nu} := C_{\nu} - \delta_{\nu} \quad (\nu=1,\ldots,n)$$

is the **lateness** and $\delta_{\nu} \geq 0$ is a prescribed due date of job ν. As we will see later, Lawler's algorithm can be generalized to stochastic scheduling problems with certain GERT precedence constraints. If the GERT network is an acyclic admissible EOR network and the objective function is the maximum expected lateness, we again obtain a polynomial algorithm.

5.2 Stochastic Single–Machine Scheduling with GERT Precedence Constraints: Basic Concepts

Suppose there is a single resource required to carry out each activity of a project described by an admissible GERT network.[18] We assume that the resource has capacity 1, which implies that at most one activity can be executed at a time. Thus, the activities which are carried out during any project realization have to be carried out one after another. This generally causes time delays for most activities, that is, in contrast to "pure" time planning of projects dealt with in chapters 2, 3 and 4, it is no longer true that each activity is begun at its earliest possible start time. We then seek to determine a sequence of activity executions that minimizes a certain objective function (cost function) subject to precedence constraints given by the GERT network. Note that we have a genuine optimization problem only if the GERT network N in question contains activities that can be carried out simultaneously. Thus, if N has only one source, N must possess at least one deterministic node.

If we think of the resource of capacity 1 as a machine and identify the activities with jobs processed on that machine, we get a stochastic single–machine scheduling problem with precedence constraints given by the admissible GERT network under consideration. Note that the activities or jobs now correspond to the arcs of the network and not to the nodes of a directed graph as in section 5.1.

---

[18]   Obviously, dummy activities have zero consumption of resources. However, since a dummy activity also has zero duration and incurs zero cost (and thus does not contribute to the objective function), we may assume that the respective resource is required to carry out the dummy activities, too.

(5.2.1) We restrict ourselves to the case where the activities of the project in question are carried out without interruption and without idle times between them. The latter assumption does not mean any loss of generality if the objective function is nondecreasing with respect to the completion times of the individual activitites (cf. Baker (1974), section 2.2). We also assume without loss of generality that the admissible GERT network under consideration has only one source (which is activated at time 0). If the original GERT network N has several sources, we consider the corresponding one–source network Ñ instead (cf. (1.3.4)) where the costs of carrying out the "additional arcs" in Ñ are zero.

(5.2.2) As stated in section 5.1, a list schedule represents a decision rule that specifies which job is to be executed next. To apply that concept to scheduling problems with GERT precedence constraints, activity executions, which occur only with a certain probability, cannot take the role of the jobs. Instead, we will introduce so–called operations, which represent deterministic quantities and give rise to the definition of nonrandomized scheduling policies. A **scheduling policy** specifies which operation is to be performed next given the project evolution up to the present. A scheduling policy is said to be a **Markov scheduling policy** if the operation to be performed next depends only on the current project state but not on the past history (cf. (1.3.5)). For the special scheduling problems with EOR precedence constraints treated in sections 5.5 and 5.7, priority lists of operations will be considered, which represent special Markov scheduling policies and are again called **list schedules**. Scheduling policies will be discussed in more detail in sections 5.8 and 5.10.

(5.2.3) In short, an **operation** is a set of activities with one and the same initial node, say node i, such that exactly one of these activities is carried out after each activation of node i. This single activity execution represents a **performance of the operation** given the project realization in question.

(5.2.4) Consider a node i different from a sink whose set of successors is $S(i)=\{j_1,\ldots,j_s\}$. If node i is stochastic, then the set of all outgoing arcs, $\{<i,j_1>,\ldots,<i,j_s>\}$, represents a **stochastic operation**. Node i is said to be the **beginning event** and $j_1,\ldots,j_s$ are the **terminal events** of the operation. A performance of a stochastic operation o consists of the execution of that activity from o which is carried out after the respective occurrence of its beginning event i during the project realization in question. If node i is deterministic, then each individual outgoing arc $\langle i, j_\sigma \rangle$ or, respectively, the corresponding one–element set $\{\langle i, j_\sigma \rangle\}$ is regarded as a **deterministic operation** with **beginning event** i and **terminal event** j_σ ($1 \leq \sigma \leq s$). The performance of that operation consists of the execution of activity $\langle i, j_\sigma \rangle$. If node i is a sink, then no operation is associated with i. The set of all operations of the admissible GERT network under consideration is denoted bei $\mathcal{O}$.

(5.2.5) Given a project realization, an operation o is called **performable** if the beginning event of o has occurred but o has not been performed yet (in other words, o is ready to be performed). If the beginning event of operation o belongs to a cycle (and o can thus be performed several times), then o is said to be performable when its beginning event has occurred β times and o has been performed $\beta-1$ times ($\beta \in \mathbb{N}$).

(5.2.6) An operation o ist said to be a **predecessor** of an operation o' if one of the terminal events of o coincides with the beginning event of o'. As usual, if o is a predecessor of o', then o' is called a **successor** of o. In Fig. 5.2.1 the stochastic operations $\{\langle 1, 3 \rangle, \langle 1, 4 \rangle\}$ and $\{\langle 2, 4 \rangle\}$ are predecessors of the deterministic operations $\{\langle 4, 5 \rangle\}$ and $\{\langle 4, 6 \rangle\}$.

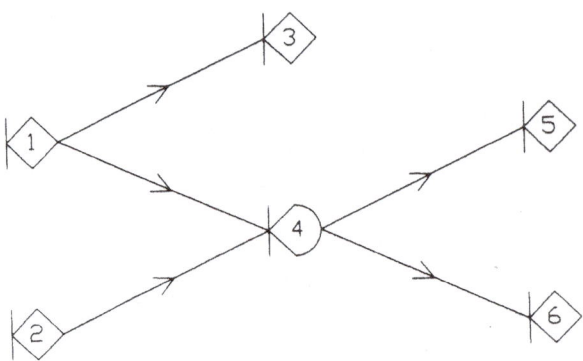

Figure 5.2.1

(5.2.7) Recall that if the precedence constraints for the jobs in a deterministic scheduling problem are given by an outtree where the jobs correspond to the nodes, each job has at most one predecessor. In an admissible acyclic EOR network N, an operation may have more than one predecessor (compare Fig. 5.2.1). During a single realization of N, however, at most one predecessor of any operation is performed. This results from the tree–structure property of N (cf. proposition (3.3.4)) or from condition (3.3.1), respectively, as follows: If an operation o has more than one predecessor, then there are at least two distinct walks in N that "separate" at some node i and "meet" at the beginning event j of operation o. Because of condition (3.3.1) node i is stochastic. Thus, at most one of these walks is executed during a single realization of N.

In the algorithms for solving scheduling problems discussed later, we have to compute certain weighting quantities of operations such as the expected duration and expected lateness. For this we establish the following

(5.2.8)   **Convention.**
          For each stochastic node i with $|S(i)|>1$, the outcome of the random experiment that selects the activity with initial node i to be carried out after an activation of node i is not known until the respective execution of the activity selected has been begun.

In convention (5.2.8) "has been begun" can be replaced by "has been terminated" because the activities of the project in question are carried out without interruption. If the outcome of the random experiment specified in convention (5.2.8) were known already before starting the respective activity execution, we would have to use the weighting quantities of the activity with initial node i selected instead of those of the stochastic operation with beginning event i.

Convention (5.2.8) can always be satisfied, if necessary, by introducing additional activities and nodes. An example is shown in Fig. 5.2.2 where the activity to be carried out after activation of node i is supposed to be known at the time of termination of activity $<k,i>$. Activity $<k,i_1>$ (or $<k,i_2>$, respectively) corresponds to activity $<k,i>$ if activity $<i,j_1>$ (or $<i,j_2>$, respectively) follows $<k,i>$. The original and the modified GERT network are equivalent (cf. definition (2.2.17)).

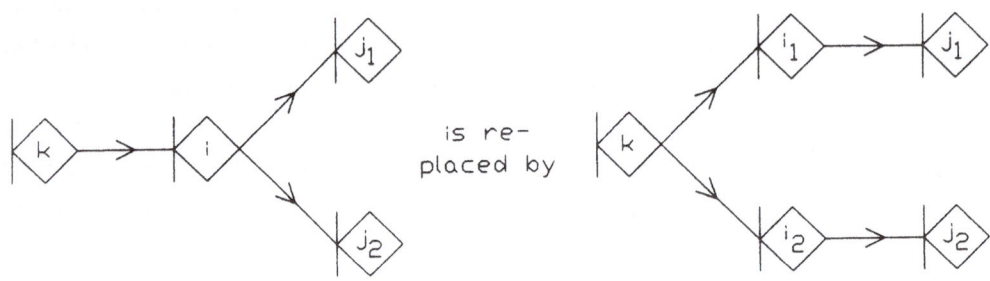

Figure 5.2.2

5.3 Stochastic Single–Machine Scheduling with GERT Precedence Constraints: Optimality Criteria and Complexity

In what follows, we consider several objective functions to be minimized for scheduling problems with precedence constraints given by an admissible GERT network. Let C_{ij}^{α} be the completion time of the αth execution of activity $<i,j>$ ($\alpha \in \mathbb{N}$), where $C_{ij}^{\alpha} := 0$ if activity $<i,j>$ is carried out less than α times. Let the function $f_{ij} : \mathbb{R}_+ \to \mathbb{R}$ be nondecreasing on $\mathbb{R}_{++}$, let $f_{ij}(0) := 0$, and let f_{ij} be bounded below and majorized by an affine function on $\mathbb{R}_+$. $f_{ij}(t)$ can be interpreted as the cost incurred when an execution of activity $<i,j>$ is completed at time t. Then the first objective function we want to minimize is

$$(5.3.1) \qquad \mathbb{E}\Big[\sum_{<i,j>\in E} \sum_{\alpha=1}^{\infty} f_{ij}(C_{ij}^{\alpha})\Big] =: \mathbb{E}(\Sigma f)$$

where $\mathbb{E}[K]$ is again the expectation of random variable K and the sum is extended over all executions of activity $<i,j>$ and all activities $<i,j>$ of the project in question. (5.3.1) represents the **expected total project cost** which is finite because the functions f_{ij} are majorized by affine functions, and because the number of activity executions and the total time of carrying out all activities that are executed during a single project execution are finite with probability 1 (cf. theorem (1.3.9)).

A special case of objective function (5.3.1) is the **expected weighted flow-time**

$$(5.3.2) \qquad \mathbb{E}[\sum_{<i,j>\in E} \sum_{\alpha=1}^{\infty} w_{ij} C_{ij}^{\alpha}] =: \mathbb{E}(\Sigma w C)$$

where $w_{ij} \geq 0$ is a weighting factor describing the importance of activity $<i,j>$. If we choose $f_{ij}(t):=d_{ij}$, where $d_{ij}:=\mathbb{E}(D_{ij})$ is the expected duration of activity $<i,j>$, then the objective function (5.3.1) is denoted by $\mathbb{E}(\Sigma d)$ and represents the **expected project duration**.

(5.3.3) Objective function (5.3.1) can be written in a different way. Let C_o^{β} be the completion time of the βth performance of opertion o ($\beta\in\mathbb{N}$), where again $C_o^{\beta}:=0$ if operation o is performed less than β times. Let $f_o^{\beta}(t)$ be the cost incurred when the βth performance of operation o is completed at time t, where f_o^{β} is supposed to have the same properties as f_{ij}. The functions f_{ij} and f_o^{β} are related to each other as follows: For a stochastic node i, $f_{ij}(C_{ij}^{\alpha})=f_o^{\beta}(C_o^{\beta})$ if $<i,j>\in o$ and the βth performance of operation o coincides with the αth execution of activity $<i,j>$. For a deterministic node i, $f_{ij}(C_{ij}^{\alpha})=f_o^{\beta}(C_o^{\beta})$ if $o=\{<i,j>\}$ and $\alpha=\beta$. In the latter case, only $\alpha=\beta=1$ is of interest because a deterministic node is activated at most once during any project realization. Thus, objective function (5.3.1) can also be written as

$$\mathbb{E}[\sum_{o\in\mathcal{O}} \sum_{\beta=1}^{\infty} f_o^{\beta}(C_o^{\beta})]$$

where $\mathcal{O}$ is again the set of all operations of the underlying GERT network.

The second type of objective function we wish to minimize is the maximum expected completion cost of activities carried out. To make precise what is to be understood by that optimality criterion, we first consider the simple example of a project shown in Fig. 5.3.1. Assume that for the two activities $<i,j_\sigma>$ ($\sigma=1,2$), the duration is $D_{ij_\sigma}=1$ and a prescribed due date $\delta_{ij_\sigma}=2$ is given. The completion cost $f_{ij_\sigma}(C_{ij_\sigma})$ of activity

$\langle i,j_\sigma \rangle$ is to be its lateness L_{ij_σ} :

$$f_{ij_\sigma}(C_{ij_\sigma}) = L_{ij_\sigma} := C_{ij_\sigma} - \delta_{ij_\sigma} \quad (\sigma=1,2)$$

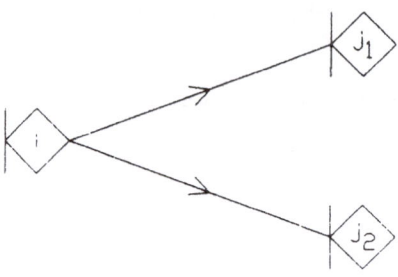

Figure 5.3.1

If the project realization consists of the execution of activity $\langle i,j_1 \rangle$, then

$$f_{ij_1}(C_{ij_1}) = -1$$
$$f_{ij_2}(C_{ij_2}) = 0$$

(recall that $C_{ij}:=0$ if activity $\langle i,j \rangle$ is not carried out and that $f_{ij}(0):=0$). If activity $\langle i,j_2 \rangle$ is carried out, then $f_{ij_1}(C_{ij_1})=0$ and $f_{ij_2}(C_{ij_2})=-1$. In any case we have

$$\max_{\langle i,j \rangle \in E} f_{ij}(C_{ij}) = 0$$

However, the maximum expected lateness of activities carried out equals -1. We obtain the latter result automatically if we consider the maximum expected lateness of the operations where the only (stochastic) operation of the project in question is $o=\{\langle i,j_1 \rangle, \langle i,j_2 \rangle\}$ (which is performed only once). By (5.3.3) we have

$$f_o^1(C_o^1) = -1$$

Hence it is expedient to define the optimality criterion in terms of the functions f_o^β $(o \in \mathcal{O})$ from the outset as follows: Let M_o be the number of performances of operation o

during a single project execution. Then the completion cost of operation o is supposed
to be

$$(5.3.4) \qquad f_o(C_o^{M_o}) := \max_{\beta=1,\ldots,M_o} f_o^{\beta}(C_o^{\beta})$$

where $\max_{\beta \in \emptyset} g_{\beta} := 0$. Then the **maximum expected completion cost of the operations** is

$$(5.3.5) \qquad \max_{o \in O} \mathbb{E}[\max_{\beta=1,\ldots,M_o} f_o^{\beta}(C_o^{\beta})] = \max_{o \in O} \mathbb{E}[f_o(C_o^{M_o})] =: \max \mathbb{E}(f)$$

Trivially, the cost (5.3.5) is finite. If the GERT network in question is acyclic and thus
each operation o is performed at most once, the objective function (5.3.5) reduces to

$$(5.3.6) \qquad \max_{o \in O} \mathbb{E}[f_o(C_o)]$$

where C_o is the completion time of operation o and $C_o := 0$ if operation o is not
performed (for simplicity, we have omitted the superscript 1 and have set $f_o := f_o^1$).

Next, we consider the special case

$$(5.3.7) \qquad f_o^{\beta}(C_o^{\beta}) = L_o^{\beta} := C_{ij}^{\alpha} - \delta_{ij} \quad \text{if the } \beta\text{th performance of operation o}$$
coincides with the αth execution of activity
$\langle i,j \rangle$

where L_o^{β} is the lateness of the βth performance of operation o and $\delta_{ij} \geq 0$ is a prescribed
due date for activity $\langle i,j \rangle$. The **maximum expected lateness of the operations** is then

$$(5.3.8) \qquad \max_{o \in O} \mathbb{E}[\max_{\beta=1,\ldots,M_o} L_o^{\beta}] =: \max \mathbb{E}(L)$$

If the underlying GERT network is acyclic, the superscript β can again be omitted and
the objective function reduces to

(5.3.9) $\max_{o \in O} \mathbb{E}(L_o)$

Note that L_o^β need not be nondecreasing with respect to β because the βth performance of operation o may represent an execution of an activity $<i,j>$ and the $(\beta+1)$st performance of operation o may coincide with an execution of a different activity $<i,k>$ whose lateness is smaller. However, the objective functions $\mathbb{E}(\Sigma f)$ and $\max \mathbb{E}(f)$ are nondecreasing with respect to the completion times of the individual activity executions or respectively operation performances.

If in (5.3.7) $\delta_{ij}=0$ for all $<i,j>\in E$, then the objective function (5.3.8) is denoted by $\max \mathbb{E}(C)$ and represents the **maximum expected completion time of the operations.** From the definition of the expected value it follows that

(5.3.10) $\max_{o \in O} \mathbb{E}(C_o) \leq \mathbb{E}(\max_{o \in O} C_o)$

where $\mathbb{E}(\max C)=\mathbb{E}(\Sigma d)$ is the expected project duration.

Scheduling problems with objective functions of type $\mathbb{E}(\Sigma f)$ are called **min–sum problems** whereas scheduling problems with objective functions of type $\max \mathbb{E}(f)$ are termed **min–max problems.** It is obvious that min–max scheduling problems for cyclic GERT networks are very hard because expected values of maxima of random variables have to be determined (cf. (5.3.5)), which requires the consideration of all possible project realizations. We will see later that for acyclic GERT networks, too, min–max scheduling problems are generally harder than the corresponding min–sum problems.

The **notation** for different types of scheduling problems with GERT precedence constraints corresponds to that one for deterministic scheduling problems (cf. section 5.1). For example, the symbol $1|\text{GERT},D\text{-}G|\mathbb{E}(\Sigma f)$ stands for a single–machine scheduling problem with objective function (5.3.1). The precedence constraints are given by an admissible GERT network (which may contain cycles) and "D~G" refers to general distribution of activity durations (recall that there is no restriction on the type of distribution of the activity durations except that the activity durations $D_{ij}, <i,j>\in E$, are supposed to be conditionally independent random variables in the sense of assumption A2).

If the underlying admissible GERT network is acyclic, "acycl" is added before "GERT", and if the GERT network represents an EOR network, "GERT" is replaced by "EOR". For instance, $1 \mid \text{acyclEOR}, D\sim G \mid \max \mathbb{E}(L)$ stands for a single–machine scheduling problem where the maximum expected lateness of the operations (5.3.8) is to be minimized and where the precedence constraints are given by an admissible acyclic EOR network. In the case of skipping (that is, the execution of activities is discontinued when no more sink can be activated, compare (1.3.13)), "skip" is added before "D~G". Note that because for admissible EOR networks, the skipping project duration equals the nonskipping project duration with probability 1 (cf. theorem (3.4.10)), we need only consider "nonskipping scheduling problems" if the precedence constraints are given by an admissible EOR network.

Finally, we discuss the **time complexity** of scheduling problems with GERT precedence constraints. First we consider min–sum problems. Problem $1 \mid \text{GERT}, D\sim G \mid \mathbb{E}(\Sigma w)$, where the objective function is the expected sum of the weights of the activities $\langle i,j \rangle$ that are carried out (note that an activity is again counted as often as it is executed), is trivial because any feasible sequence of activity executions [19] is optimal. That trivial problem also covers the case where we wish to minimize the expected project duration $\mathbb{E}(\Sigma d)$. On the other hand, even the scheduling problems $1 \mid \text{prec}, D_\nu = 1 \mid \Sigma w_\nu C_\nu$ and $1 \mid \text{prec}, D_\nu \sim \exp(1) \mid \mathbb{E}(\Sigma w_\nu C_\nu)$ with precedence constraints for the jobs ν given by an acyclic directed graph, (expected) weighted flow–time as objective function, and unit–time jobs or respectively exponentially distributed job durations are NP–hard (cf. Lenstra and Rinnooy Kan (1978), Pinedo (1982)). Hence, the more general scheduling problems $1 \mid \text{acyclGERT}, D=1 \mid \mathbb{E}(\Sigma wC)$ and $1 \mid \text{acyclGERT}, D\sim G \mid \mathbb{E}(\Sigma wC)$ are NP–hard, too. The special problem $1 \mid \text{EOR}, D\sim G \mid \mathbb{E}(\Sigma wC)$, however, can be solved in polynomial time as we will see in section 5.5.

As already mentioned, min–max scheduling problems are generally harder than min–sum problems. In particular, we have

(5.3.11)   **Theorem.**
            Problem $1 \mid \text{acyclEOR}, D=1 \mid \max \mathbb{E}(C)$ is NP–hard.

---

[19]   "feasible" refers to the capacity and precedence constraints

Proof (cf. Bücker (1990)).

We consider the following version of the NP–hard 0–1 knapsack problem (compare Garey and Johnson (1979), sections 3.2.1 and A6, and Papadimitriou and Steiglitz (1982), section 15.7): Given $g_1, \ldots, g_n, g \in \mathbb{R}_{++}$ where $n > 1$ and $g < \sum_{j=1}^{n} g_j$. Find $J \subseteq \{1, \ldots, n\}$ such that $|\sum_{j \in J} g_j - g|$ is minimum. Obviously, the following similar problem (P) is also NP–hard.

(P) Given $p_1, \ldots, p_n$ with $0 < p_j < 1$ for $j = 1, \ldots, n > 1$ and $\sum_{j=1}^{n} p_j = 1$.

   Find $J \subseteq \{1, \ldots, n\}$ such that $|\sum_{j \in J} p_j - \frac{1}{2}|$ is minimum.

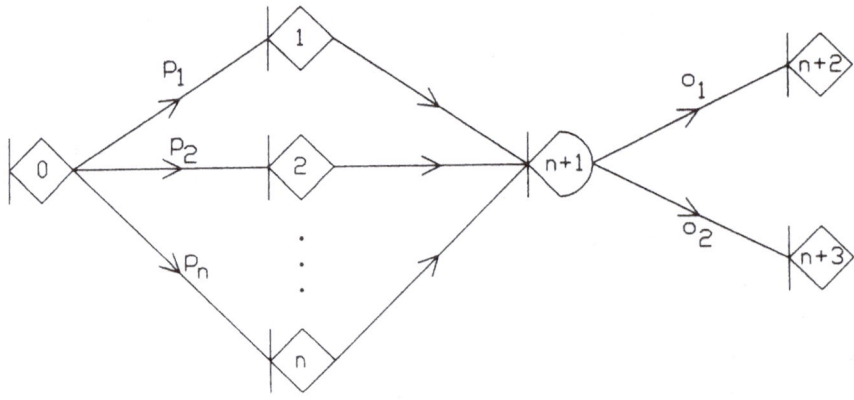

Figure 5.3.2

We show that Problem (P) is equivalent to the scheduling problem $1 | \text{acyclEOR}, D=1 | \max \mathbb{E}(C)$ for the EOR network of Fig. 5.3.2, where the activities with initial node 0 are marked with their execution probabilities and all activity durations are equal to 1. The project evolution up to the activation of node $n+1$ is uniquely specified by the path $\langle 0, j, n+1 \rangle$ executed or, more briefly, by the node j activated ($1 \leq j \leq n$). We express this fact by saying that, when node $n+1$ is activated, the project is in one of the "states" j ($1 \leq j \leq n$). The performable operations in each "state" $j \in \{1, \ldots, n\}$ are $o_1 = \{\langle n+1, n+2 \rangle\}$ and $o_2 = \{\langle n+1, n+3 \rangle\}$. Thus, each scheduling policy can be identified with a mapping $\phi: \{1, \ldots, n\} \to \{o_1, o_2\}$, where $\phi(j) = o_\nu$ means that operation o_ν is performed first after execution of path $<0,j,n+1>$ $(\nu=1,2;1\leq j\leq n)$.

Let

$$J(\phi) := \{j\,|\,\phi(j)=o_1\}$$

be the set of "states" j for which operation o_1 is performed first and let

$$p(\phi) := \sum_{j\in J(\phi)} p_j$$

be the probability that operation o_1 is performed first when scheduling policy ϕ is applied. Moreover, let $C_o(\phi)$ be the completion time of operation o for policy ϕ. Then

$$\mathbb{E}[C_{o_1}(\phi)] = 3p(\phi)+4(1-p(\phi)) = 4-p(\phi)$$
$$\mathbb{E}[C_{o_2}(\phi)] = 4p(\phi)+3(1-p(\phi)) = 3+p(\phi)$$

and the value of the objective function for policy ϕ becomes

$$\max_{o\in O} \mathbb{E}[C_o(\phi)] = \max(\mathbb{E}[C_{o_1}(\phi)],\mathbb{E}[C_{o_2}(\phi)])$$
$$= \max(4-p(\phi),3+p(\phi))$$

Since $4-p(\phi)$ is decreasing and $3+p(\phi)$ is increasing with respect to $p(\phi)$, a scheduling policy ϕ is optimal exactly if it minimizes $|p(\phi)-\frac{1}{2}|$. In other words, we have to find ϕ such that $|\sum_{j\in J(\phi)} p_j-\frac{1}{2}|$ is minimum, which corresponds to problem (P).

Obviously, for any scheduling policy ϕ, $\mathbb{E}[\max_{o\in O} C_o(\phi)]=4$ is the (expected) project duration and

$$\max_{o\in O} \mathbb{E}[C_o(\phi)] < \mathbb{E}[\max_{o\in O} C_o(\phi)]$$

if $J(\phi)\neq\emptyset,\{1,\ldots,n\}$ and thus $0<p(\phi)<1$ (compare (5.3.10)).

Note that the decision which of the two operations o_1 or o_2 is performed first after activation of node n+1 generally depends on the past history (that is, which of the paths $<0,j,n+1>$, $1 \leq j \leq n$, has been executed). Thus, as a rule, an optimal scheduling policy is not a Markov policy (cf. (5.2.2)).   ■

From theorem (5.3.11) it follows that the more complicated scheduling problem $1|\text{acyclEOR},D\text{~}G|\text{maxE}(L)$ is NP–hard, too. In Bücker (1990) it is shown that problem $1|\text{acyclEOR},D=1|\text{maxE}(C)$ is NP–hard even if we restrict ourselves to list schedules (cf. section 5.4) as feasible scheduling policies. However, if we further restrict the set of feasible solutions to so–called precedence schedules (cf. section 5.4), there is a polynomial algorithm which solves problem $1|\text{acyclEOR},D\text{~}G|\text{maxE}(L)$ and will be presented in section 5.7. The problem $1|\text{acyclGERT},D=1|\text{maxE}(L)$ is NP–hard even if we restrict ourselves to precedence schedules (cf. Bücker (1988, 1990)).

Skipping scheduling problems are in general harder than the corresponding nonskipping problems. For example, problem $1|\text{GERT},\text{skip},D=1|E(\Sigma w)$ is NP–hard (there is a polynomial reduction of the NP–hard minimum–cover problem to that scheduling problem, cf. Rubach (1984)) in contrast to the trivial problem $1|\text{GERT},D\text{~}G|E(\Sigma w)$.

5.4 List Schedules and Sequences of Activity Executions

(5.4.1) Let $\mathcal{O}$ be again the set of all operations of the GERT network under consideration and let $m:=|\mathcal{O}|$. A sequence of operations $(o_{\nu_1},\ldots,o_{\nu_m})$ where $\nu_i \neq \nu_j$ for $i \neq j$ ($1 \leq i,j \leq m$) is called a **list schedule**. The **interpretation of a list schedule** $\mathbb{Q}$ is as follows. The operation to be performed next is always that performable operation (cf. (5.2.5)) which has the foremost position in list schedule $\mathbb{Q}$ among all performable operations. Thus, a list schedule represents a priority list of operations. List schedules are special Markov scheduling policies (cf. (5.2.2)) where each operation has a fixed priority (its position in the list schedule) independent of the past history. If in a list schedule $\mathbb{Q}=(o_{\nu_1},\ldots,o_{\nu_m})$, we have i<j (that is, o_{ν_i} is "before" o_{ν_j}) whenever o_{ν_i} is a predecessor of o_{ν_j} ($1 \leq i < j \leq m$), then $\mathbb{Q}$ is said to be a **precedence schedule** (because it takes the precedence constraints into account). Obviously, precedence schedules make sense only when the underlying GERT network N is acyclic. If N is an acyclic EOR network, the sequence of operations performed during any realization of N always represents a subsequence of a precedence schedule. A list schedule or precedence schedule that minimizes the objective function of the scheduling problem in question (with respect to all list schedules or precedence schedules, respectively) is said to be **optimal**.

Since the performance of any operation corresponds to precisely one activity execution, a unique sequence of activity executions is specified given a list schedule and a project realization. The preceding interpretation of a list schedule immediately provides an algorithm that finds the sequence of activity executions L associated with a given list schedule Q and project realization ω. The activity sequence L and list schedule Q are supposed to be implemented as linear lists (for the data structures list and priority queue used subsequently we refer to Aho et al. (1983), sections 2.1 and 4.10). The project realization ω is supposed to be given by a mapping $h : O_{st} \times \mathbb{N} \rightarrow E$, where O_{st} is the set of the stochastic operations, with the following meaning: $h(o,\beta) = \langle i,j \rangle$ says that the βth performance of stochastic operation o (which occurs after the βth occurrence of the beginning event i of o) represents an execution of activity $\langle i,j \rangle$. The set of performable operations is to be implemented as a priority queue P where the priority of an operation o in P is equal to the position of o in list schedule Q (that is, the first operation in Q has the smallest priority 1 and the last operation in Q has priority $|Q|=m$). So, we have

(5.4.2)   Algorithm.

Step 1.   Set $P:=\emptyset, L:=\emptyset$ and $\beta_o:=0$ for every stochastic operation o
Insert all operations into P whose beginning event is the source of the network

Step 2.   Delete operation o with smallest priority from P
If o is deterministic, let $o=\{\langle i,j \rangle\}$;
otherwise set $\beta_o:=\beta_o+1$ and let $\langle i,j \rangle = h(o,\beta_o)$
Insert $\langle i,j \rangle$ at the rear of L
Insert all operations with beginning event j into P

Step 3.   If $P=\emptyset$, then terminate (L contains the sequence of activities carried out); otherwise go to step 2   ∎

Let k be the number of activity executions in project realization ω. Since, for a priority queue of m elements, the operations "insert" and "delete" take $O(\log m)$ time (cf. Aho et al. (1983), section 4.11), algorithm (5.4.2) can be implemented to run in $O(k \log m)$ time.

As an example we consider the admissible EOR network (without arc weights) shown in Fig. 5.4.1. The first subscript of a deterministic operation coincides with the number of its beginning event and the second is its serial number. The activities carried out during the project realization in question are illustrated by darker arrows. Assume that if node 6 is activated for the first time, activity <6,4> is executed, and if node 6 is activated for the second time, activity <6,9> is carried out (in other words, at first we run through the cycle <4,6,4> and when we arrive at node 6 for the second time, activity <6,9> will be begun).

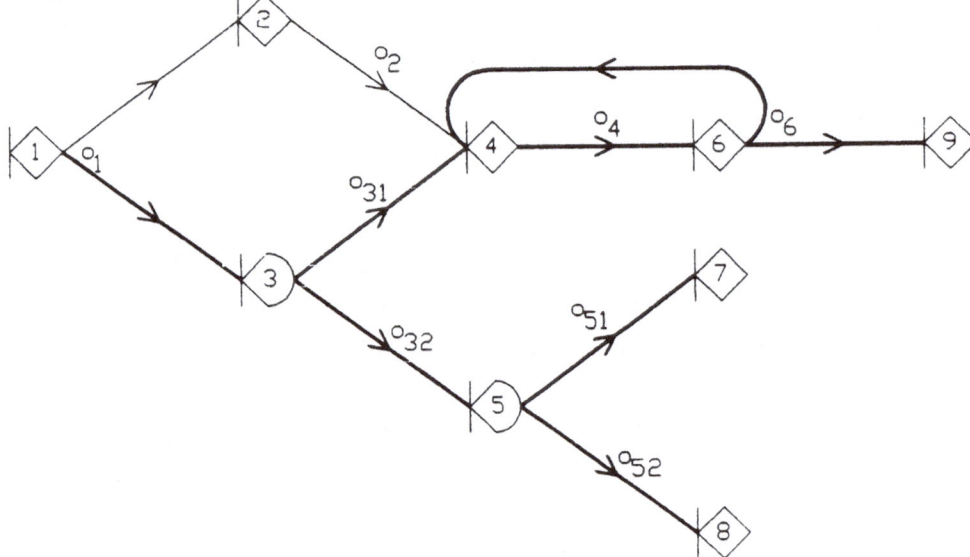

Figure 5.4.1

Event occurred	Performable operations	Operation performed	Activity carried out
1	(o_1)	o_1	<1,3>
3	(o_{32}, o_{31})	o_{32}	<3,5>
5	(o_{52}, o_{31}, o_{51})	o_{52}	<5,8>
8	(o_{31}, o_{51})	o_{31}	<3,4>
4	(o_4, o_{51})	o_4	<4,6>
6	(o_6, o_{51})	o_6	<6,4>
4	(o_4, o_{51})	o_4	<4,6>
6	(o_6, o_{51})	o_6	<6,9>
9	(o_{51})	o_{51}	<5,7>

Table 5.4.1

Suppose the list schedule $\mathbf{Q}=(o_{32},o_2,o_6,o_1,o_{52},o_4,o_{31},o_{51})$ is given. Then Table 5.4.1 shows the current performable operations (ordered according to increasing priorities) as well as the sequences of events occurred, of operations performed and of activities carried out. The second column in Table 5.4.1 gives the current priority queue P and the last column shows the list of activity executions L.

5.5 Minimum Flow–Time Scheduling in EOR Networks

In this section we deal with the scheduling problem $1|EOR,D\text{-}G|\mathbb{E}(\Sigma wC)$ where the precedence constraints are given by an admissible EOR network with at least one deterministic node (most of the following material is taken from Bücker and Neumann (1989) and Bücker et al. (1990)). Exploiting the tree–structure property of an admissible EOR network (cf. proposition (3.3.4)) we present a polynomial algorithm, which computes an optimal list schedule and is a generalization of the procedure for the deterministic scheduling problem $1|tree|\Sigma w_\nu C_\nu$ (compare (5.1.3)). In Bücker (1990) is shown that for problem $1|EOR,D\text{-}G|\mathbb{E}(\Sigma wC)$, an optimal list schedule also represents an optimal scheduling policy, that is, an optimal solution. How to find an optimal sequence of activity executions given a project realization and an optimal list schedule has been shown in section 5.4.

Bruno (1976) has dealt with a stochastic single–machine scheduling problem with tree precedence constraints. A set of jobs without precedence constraints is to be processed on a single machine. Each job is considered a decision tree whose nodes correspond to tasks. The execution time of a task and the decision as to which of the following tasks, if any, is executed next are stochastic. The execution of a job consists of the execution of some chain of tasks in the tree, where a job execution may be interrupted only at task terminations in order to assign the machine to another job. The objective function is the expected weighted sum of the task completion times.

Bruno's scheduling problem can be viewed as a problem of type $1|acyclEOR,D\text{-}G|\mathbb{E}(\Sigma wC)$ for a special acyclic EOR network whose arcs correspond to the tasks. The general problem $1|EOR,D\text{-}G|\mathbb{E}(\Sigma wC)$ additionally permits precedence constraints for the jobs and feedback to be taken into account.

The generalization of the algorithm from (5.1.3) to problem $1|EOR,D\text{-}G|\mathbb{E}(\Sigma wC)$ is as follows. Let $d_{ij}:=\mathbb{E}(D_{ij})$ be again the expected duration of activity $<i,j>$, let

$$\tau_\nu := \sum_{<i,j>\epsilon o_\nu} p_{ij}d_{ij}$$

be the expected duration and

$$v_\nu := \sum_{<i,j>\epsilon o_\nu} p_{ij}w_{ij}$$

be the expected weight of operation o_ν, and let

$$\rho_\nu := \frac{\tau_\nu}{v_\nu} \quad \text{where} \quad \rho_\nu := \begin{cases} 0 \text{ if } \tau_\nu=0 \\ \infty \text{ if } \tau_\nu>0 \end{cases} \quad \text{for } v_\nu=0$$

Let ν be the beginning event and S_ν be the set of the terminal events of operation o_ν. Moreover, let O be again the set of operations and m be the number of operations of the underlying EOR network, and let Q be the current list schedule to be implemented as a linear list. Then we have the following algorithm that computes an optimal list schedule:

(5.5.1)   Algorithm.

Step 1.   Set $Q:=\emptyset$ and $\mu:=m$

Compute ρ_ν for all $o_\nu \epsilon O$

Step 2.   Find $o_1 \epsilon O$ such that $\rho_1 = \min_{o_\nu \epsilon O} \rho_\nu$

Insert o_1 at the rear of Q, delete o_1 from O, and set $\mu:=\mu-1$

Step 3.   If $\mu=0$, then terminate (Q is an optimal list schedule)

For each predecessor $o_k \epsilon O$ of o_1 set

$$\tau_k := \frac{\tau_k+p_{k1}\tau_1}{1-p_{k1}p_{1k}}, \quad v_k := \frac{v_k+p_{k1}v_1}{1-p_{k1}p_{1k}}, \quad \rho_k := \frac{\tau_k}{v_k}$$

$$S_k := \begin{cases} S_k \cup S_1 \setminus \{k,1\} & \text{if node 1 is stochastic} \\ S_k \cup S_1 & \text{if node 1 is deterministic} \end{cases}$$

$$p_{ks} := \frac{p_{ks}+p_{k1}p_{1s}}{1-p_{k1}p_{1k}} \quad \text{for all } s\epsilon S_k$$

where $p_{ij}:=0$ if there is no arc $<i,j>$

Go to step 2   ■

The heart of each of the m iterations of algorithm (5.5.1) is that two operations o_k and o_1, where o_k is a predecessor of o_1, are replaced by one **composite operation**, which is again denoted by o_k. An example is illustrated in Fig. 5.5.1, where each activity $\langle i,j \rangle$ is marked with its execution probability p_{ij} and where we have the operations $o_k=\{\langle k,1 \rangle,\langle k,j \rangle\}$ and $o_1=\{\langle 1,k \rangle,\langle 1,i \rangle,\langle 1,j \rangle,\langle 1,r \rangle\}$. The sets of terminal events of o_k and o_1 are $S_k=\{1,i\}$ and $S_1=\{k,i,j,r\}$, respectively. Then the composite operation has the beginning event k and the set of terminal events $\{i,j,r\}$. The substitute activities $\langle k,i \rangle$, $\langle k,j \rangle$ and $\langle k,r \rangle$ in the composite operation have the execution probabilities

$$\frac{p_{ki}+p_{k1}p_{1i}}{1-p_{k1}p_{1k}}=\frac{7}{12}\ ,\quad \frac{p_{k1}p_{1j}}{1-p_{k1}p_{1k}}=\frac{1}{6}\quad \text{and}\quad \frac{p_{k1}p_{1r}}{1-p_{k1}p_{1k}}=\frac{1}{4}\ ,\ \text{respectively.}$$

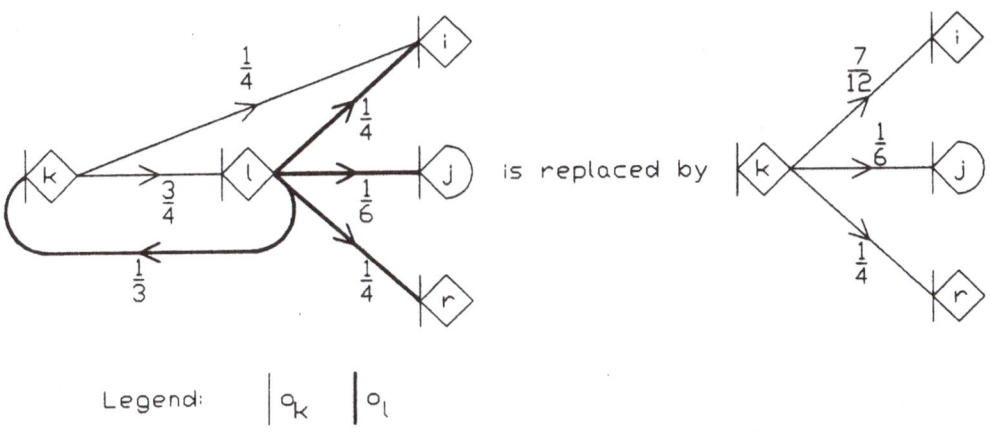

Figure 5.5.1

(5.5.2) In the course of algorithm (5.5.1), the composition of operations may result in new operations that represent neither of the two types stochastic or deterministic introduced in section 5.2. Two examples are shown in Figs. 5.5.2 and 5.5.3 where each activity is marked with its execution probability. In Fig. 5.5.2, after replacing the two operations o_k and o_1 by one composite operation, which is again designated by o_k, the beginning event k of composite operation o_k is neither deterministic nor stochastic. However, since k behaves like a deterministic node as far as the updating formula for set S_k in algorithm (5.5.1) is concerned, we call node k a **quasi–deterministic node**

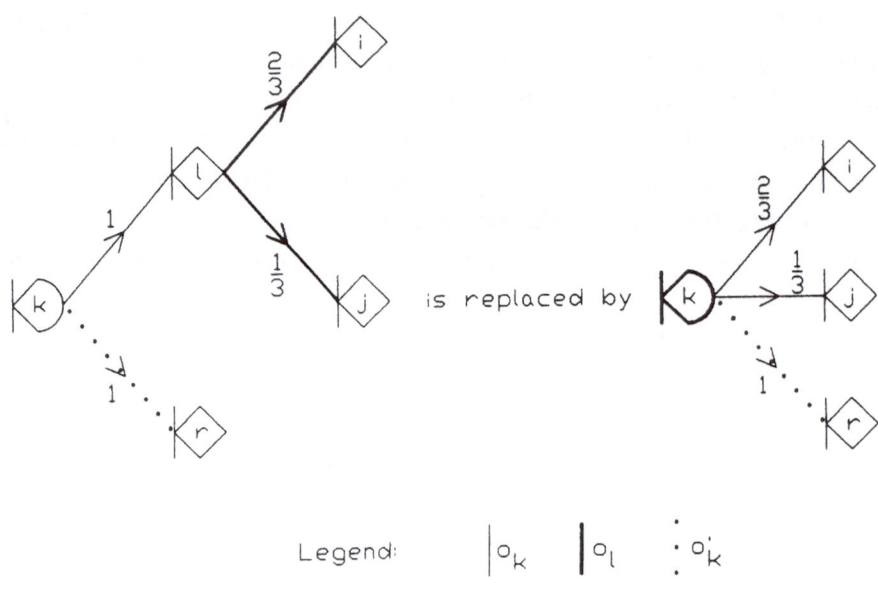

Legend: $|o_k \quad |o_l \quad \vdots o_k$

Figure 5.5.2

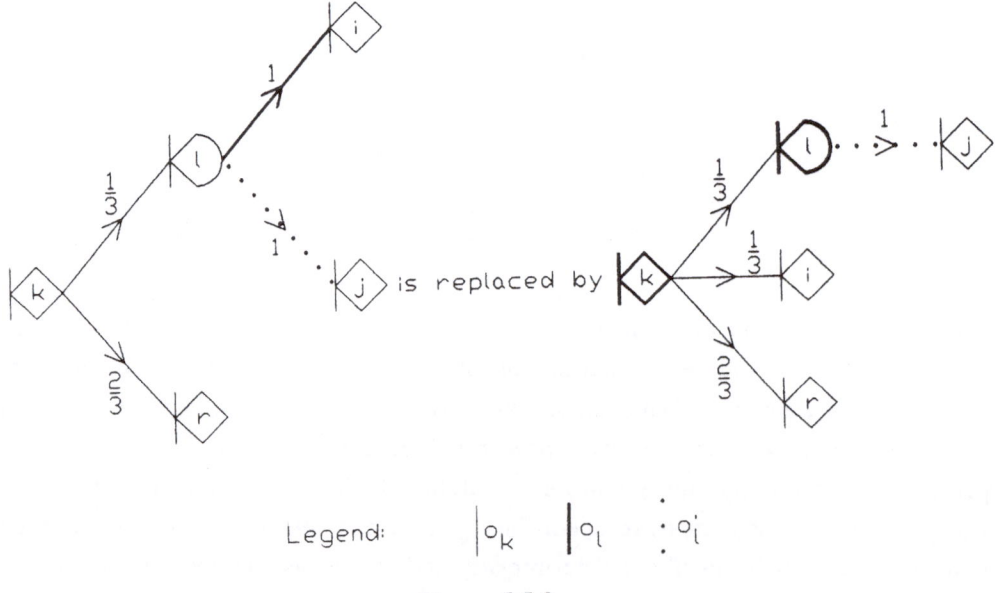

Legend: $|o_k \quad |o_l \quad \vdots o_l$

Figure 5.5.3

(represented by a darker node in Fig. 5.5.2) and continue regarding k as a deterministic node in the course of the algorithm. In Fig. 5.5.3, beginning event k of composite operation o_k behaves likes a stochastic node and is called a **quasi–stochastic node** although the sum of the execution probabilities of the outgoing activities is greater than 1. After the composition, the beginning event 1 of operation o_1 behaves like a deterministic node (and is termed a quasi–deterministic node) even though it has the single successor j. In Fig. 5.5.3 the quasi–stochastic node k and quasi–deterministic node 1 are again represented by darker nodes. In summary, each node is considered preserving its type in the course of algorithm (5.5.1).

Let again V be the node set and E be the arc set of the EOR network in question. Since $m=0(|E|)$ and the calculations to be performed in each iteration take $0(|E||V|)$ time, algorithm (5.5.1) can be implemented to run in $0(|E|^2|V|)$ time. Proofs that the algorithm produces an optimal list schedule can be found in Bücker (1990), Bücker and Neumann (1989), and Bücker et al. (1990). The proofs exploit the tree–structure property of an admissible EOR network and the Markov and independence properties from assumption A2. As already mentioned, an optimal list schedule also represents an optimal solution to problem $1|EOR,D\text{-}G|E(\Sigma wC)$ (cf. Bücker (1990)).

The quantities ρ_ν in algorithm (5.5.1) correspond to the so–called **dynamic allocation indices** or **Gittins' indices** (compare Gittins (1989)) because the (composite) operations o_ν are scheduled in nondecreasing order of the "priority indices" ρ_ν in an optimal scheduling policy. There is no way of adapting the polynomial algorithm (5.5.1) for problem $1|EOR,D\text{-}G|E(\Sigma wC)$ to scheduling problems with more general GERT precedence constraints or more general objective functions. Even the deterministic scheduling problems $1|prec|\Sigma C_\nu$ (cf. Lenstra and Rinnooy Kan (1978)) and $1||\Sigma w_\nu T_\nu$ (cf. Lenstra et al. (1977)), where $T_\nu := \max(L_\nu, 0)$ is the tardiness of job ν, are NP–hard.

The scheduling problem $1|EOR,D\text{-}G|E(\Sigma wC)$ implies that when an activity $\langle i,j \rangle$ is carried out for the αth time, a term $w_{ij}C_{ij}^\alpha$ is added to the cost of the respective project, in other words, an activity is taken into consideration for the objective function (5.3.2) as often as it is carried out. In practice, however, it is often more expedient to take into account each activity only once. To do so, we consider a progression through a cycle structure C of the underlying EOR network when C is entered at entrance node k to be one operation o_k called a **cycle operation**. For the sake of convenience, that progression is supposed to begin with the first activity execution inside C and to terminate at the first activation of a final node of any exit arc of C (not at the last activation of any exit node of C!). The performance of cycle operation o_k consists of the successive executions of those activities of C including the exit arcs of C which are carried out after the first activation of node k during the project realization in question.

To apply algorithm (5.5.1) to that **modification of problem** $1|\text{EOR},\text{D-G}|\text{E}(\Sigma w C)$, we have to compute the expected duration τ_k and an appropriate (mean) weight v_k of cycle operation o_k. For this we expand cycle structure C in a way different from that one discussed in section 4.2 (compare (4.2.1)). As in (4.2.1) we first add a STEOR node k^+ and a dummy activity $<k^+,k>$ to C, where all entrance arcs of C with final node k now lead into node k^+ and where k^+ represents the beginning event of operation o_k. Second, we add to C all exit arcs of C. The duration D_{k^+k} and weighting factor w_{k^+k} of dummy activity $<k^+,k>$ are equal to 0. For simplicity, the resulting structure C^+ is again called an **expanded cycle structure**. C^+ represents a STEOR network if it is treated as a separate entity and if each final node of an exit arc of C is considered having a stochastic exit.

Let V^+ be the set of the nodes of C^+ different from source k^+ and let E^+ be the set of the arcs of C^+ aside from dummy activity $<k^+,k>$. For $j\in V^+$ let z_{k^+j} be the activation number of node j given source k^+ in the STEOR network C^+. Recall that the activation numbers z_{k^+j} satisfy the system of linear equations

$$(5.5.3) \qquad z_{k^+j} = p_{k^+j} + \sum_{i\in V^+} p_{ij}z_{k^+i} \quad (j\in V^+)$$

where $p_{ij}:=0$ if there is no arc $<i,j>$ (compare (3.5.14)). Then the expected duration τ_k of cycle operation o_k, which equals the expected sojourn time in expanded cycle structure C^+ provided that C^+ has been entered at node k^+, is

$$(5.5.4) \qquad \tau_k = \text{E}\Big[\sum_{<i,j>\in E^+} \sum_{\alpha=1}^{M_{ij}(k^+)} D_{ij}^\alpha \Big] - \sum_{<i,j>\in E^+} z_{k^+i}p_{ij}d_{ij}$$

where D_{ij}^α is again the duration of the αth execution of activity $<i,j>$ and $M_{ij}(k^+)$ is the number of executions of activity $<i,j>$ when C^+ has been entered at source k^+. To obtain (5.5.4) assumption A2 and Wald's equation (cf. Rohatgi (1976), section 14.3) have been used.

To compute the (mean) weight v_k of cycle operation o_k, we have to replace d_{ij} by the weighting factor w_{ij} of activity $<i,j>$ in (5.5.4). Moreover, we have to recall that in the present modification of scheduling problem $1|EOR,D\text{-}G|E(\Sigma wC)$, each activity is taken into consideration for the objective function only once even if it is carried out several times. Hence, the expected number of times $z_{k^+i}p_{ij}$ activity $<i,j>$ is carried out given that source k^+ of C^+ has been activated has to be replaced by the respective execution probability $q_{k^+i}p_{ij}$ of activity $<i,j>$. Here q_{k^+i} is the conditional probability that node i is activated given that source k^+ has been activated (transition probability from k^+ to i). So,

$$(5.5.5) \qquad v_k = \sum_{<i,j> \in E^+} q_{k^+i} p_{ij} w_{ij}$$

To find q_{k^+i} we delete all arcs from C^+ which emanate from node i and solve a system of linear equations similar to (5.5.3), compare (3.5.17).

Let R^+ and S^+ be the source set and sink set, respectively, of C^+. Cycle operation o_k corresponds to the stochastic operation with beginning event k^+ and set of terminal events S^+. The execution probability of the substitute activity $<k^+,1>$, $1 \in S^+$, equals the activation number z_{k^+1}. We see that the introduction of cycle operations o_k corresponds to the replacement of C^+ by the complete bipartite STEOR network with source set R^+ and sink set S^+ (compare (4.2.2)). However, for the individual activities $<k^+,1>$ $(k^+ \in R^+, 1 \in S^+)$ of that bipartite STEOR network, we need only compute the execution probabilities but no expected durations and weighting factors.

Figs. 5.5.4 and 5.5.5 show a cycle structure C and the corresponding expanded cycle structure C^+, where each arc is marked with its execution probability (if it is less than 1) and the dummy activity $<1^+,1>$ is indicated by a broken arrow. Node 1^+ is the source and nodes 4, 5 and 6 are the sinks of C^+. Note that node 6 is regarded as a stochastic node if it is treated as a node of C^+. Cycle operation o_1 represents the progression through C or C^+ when C or C^+ is entered at node 1 or 1^+, respectively. Fig.

5.5.6 shows the stochastic operation corresponding to cycle operation o_1 and the remaining operations o_4, o_5, o_{61}, o_{62}, and o_7.

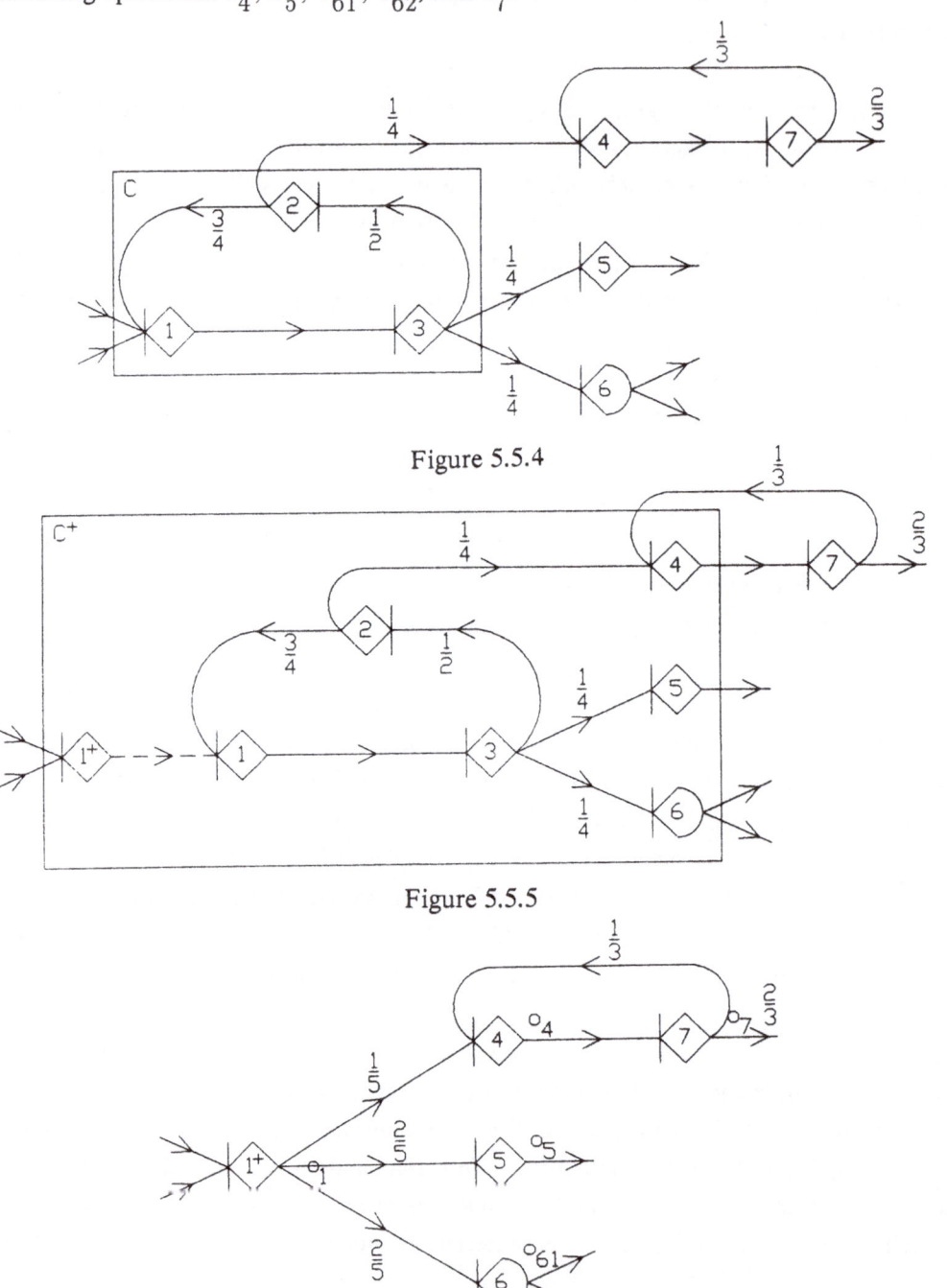

Figure 5.5.4

Figure 5.5.5

Figure 5.5.6

After the computation of τ_k and v_k for each cycle operation o_k of the underlying EOR network and the replacement of each o_k by a stochastic operation, algorithm (5.5.1) can be applied to the resulting acyclic EOR network. For each source k^+ of an expanded cycle structure, the computation of the quantities z_{k^+j} requires $0(|V|^3)$ time. For each source k^+ and node j, the computation of the probabilities q_{k^+j} takes $0(|V|^3)$ time, too, that is, altogether $0(|V|^5)$ time in the worst case (compare (3.5.18)). Since the time complexity of algorithm (5.5.1) is $0(|E|^2|V|)=0(|V|^5)$, the algorithm for solving the modification of problem $1|EOR,D\text{-}G|E(\Sigma wC)$ can also be implemented to run in $0(|V|^5)$ time.

5.6 A Flow–Time Scheduling Example

To illustrate algorithm (5.5.1) for problem $1|EOR,D\text{-}G|E(\Sigma wC)$, we consider the EOR network of Fig. 5.6.1 where each activity $<i,j>$ is marked with its execution probability p_{ij}, expected duration d_{ij}, and weighting factor w_{ij}. The darker arrows show a specific project realization where activity $<6,8>$ is carried out after the first activation of node 6 and activity $<6,7>$ is carried out when node 6 has been activated for the second time.

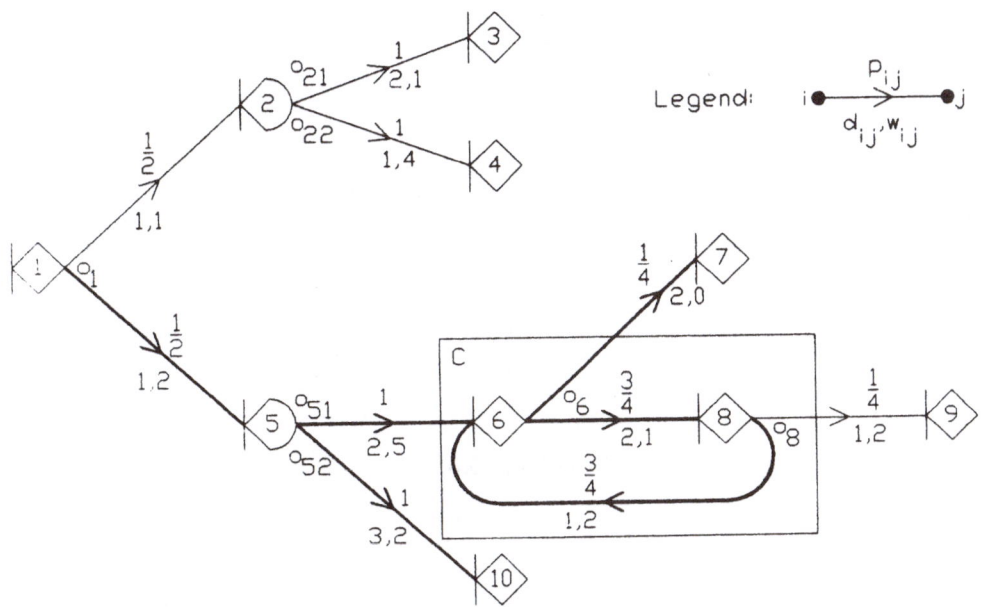

Figure 5.6.1

At the beginning we have $Q=\emptyset$ and $O=\{o_1,o_{21},o_{22},o_{51},o_{52},o_6,o_8\}$. Then the individual iterations of the algorithm are as follows where the quantities $\tau_\nu, v_\nu, \rho_\nu$, and S_ν for the operations o_ν are listed in Table 5.6.1 (the initial values of those quantities for o_1 and o_6 are given before the first comma).

o_ν	τ_ν	v_ν	ρ_ν	S_ν
o_1	1,1.5,2.5	1.5,3.5,6	0.67,0.43,0.42	$\{2,5\},\{2,4,5\},\{2,4,5,6\}$
o_{21}	2	1	2	$\{3\}$
o_{22}	1	4	0.25	$\{4\}$
o_{51}	2	5	0.4	$\{6\}$
o_{52}	3	2	1.5	$\{10\}$
o_6	2,6.29	0.75,5.14	2.67,1.22	$\{7,8\},\{7,9\}$
o_8	1	2	0.5	$\{6,9\}$

Table 5.6.1

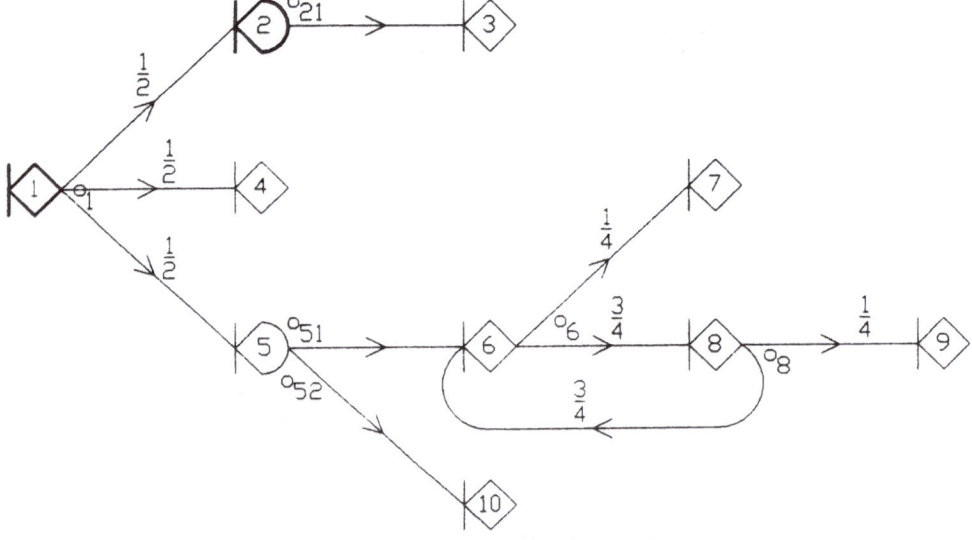

Figure 5.6.2

Iteration 1. We have

$$\rho_{22} = \min_{o_\nu \in O} \rho_\nu$$

The predecessor of o_{22} is o_1. Thus we replace the two operations o_1 and o_{22} by a composite operation, which is again denoted by o_1. For that composite operation o_1 we obtain $\tau_1=1.5$, $v_1=3.5$, $\rho_1=0.43$ and $S_1=\{2,4,5\}$ (listed in Table 5.6.1 after the first comma in the line belonging to operation o_1). At the end of iteration 1 we have $Q=(o_{22})$ and $O=\{o_1,o_{21},o_{51},o_{52},o_6,o_8\}$. The network that corresponds to the new set of unscheduled operations O is illustrated in Fig. 5.6.2 where each activity is marked with its execution probability (if it is less than 1). Note that node 1 has become quasi–stochastic and node 2 quasi–deterministic (cf. (5.5.2)) which is indicated by darker nodes in Fig. 5.6.2.

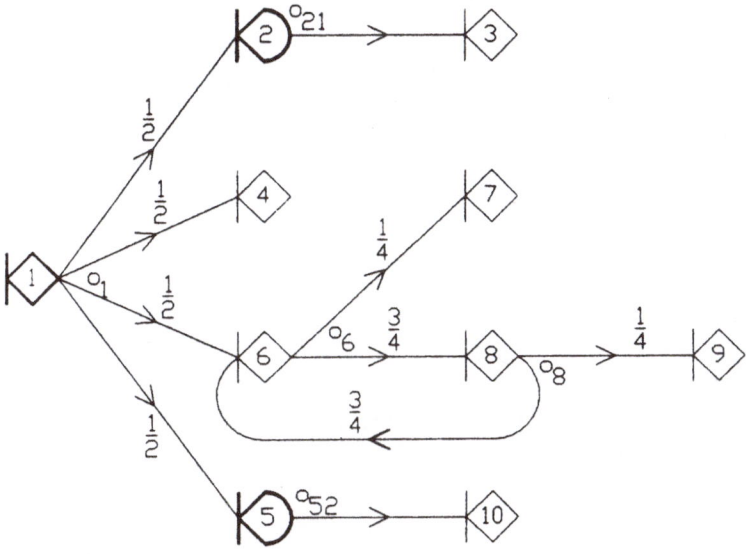

Figure 5.6.3

Iteration 2. We have

$$\rho_{51} = \min_{o_\nu \in O} \rho_\nu$$

The predecessor of o_{51} is o_1. For the composite operation o_1, which replaces the pair (o_1, o_{51}), we obtain $\tau_1 = 2.5$, $v_1 = 6$, $\rho_1 = 0.42$, and $S_1 = \{2,4,5,6\}$ (these new values are again listed in Table 5.6.1). Moreover, we get $Q = (o_{22}, o_{51})$ and $\mathcal{O} = \{o_1, o_{21}, o_{52}, o_6, o_8\}$. The network corresponding to $\mathcal{O}$ is shown in Fig. 5.6.3. Node 5 has become quasi–deterministic.

Iteration 3. Operation o_1 has minimum ρ_ν. Since o_1 has no predecessor, there is no composition of operations and we get $Q = (o_{22}, o_{51}, o_1)$ and $\mathcal{O} = \{o_{21}, o_{52}, o_6, o_8\}$. Fig. 5.6.4 shows the network belonging to $\mathcal{O}$.

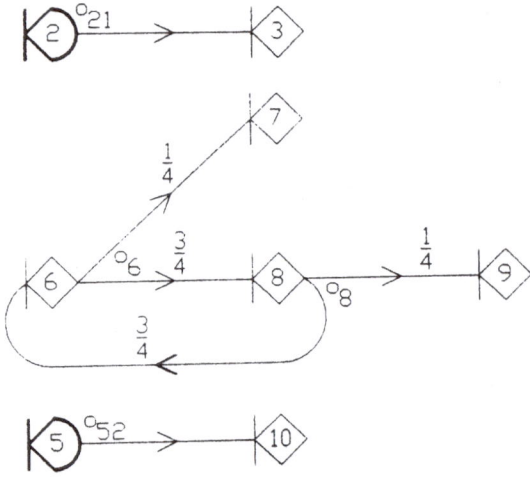

Figure 5.6.4

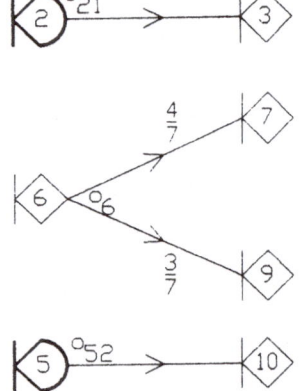

Figure 5.6.5

Iteration 4. Operation o_8 has minimum ρ_ν, its predecessor is o_6. (o_6,o_8) is replaced by the composite operation o_6 with $\tau_6=6.29$, $v_6=5.14$, $\rho_6=1.22$, and $S_6=\{7,9\}$ (compare Table 5.6.1). Furthermore, we obtain $Q=(o_{22},o_{51},o_1,o_8)$ and $O=\{o_{21},o_{52},o_6\}$. Fig. 5.6.5 shows the network corresponding to O.

In *iterations 5, 6* and 7 the operations o_6, o_{52} and o_{21}, respectively, have minimum ρ_ν and are inserted in that order at the rear of Q and deleted from O. None of those operations has any predecessor in O. Hence, no new composite operations are formed. At the end of iteration 7, we have $O=\emptyset$ and $Q=(o_{22},o_{51},o_1,o_8,o_6,o_{52},o_{21})$ is an optimal list schedule. The optimal sequence of activities carried out during the project realization indicated by darker arrows in Fig. 5.6.1 is $(\langle1,5\rangle,\langle5,6\rangle,\langle6,8\rangle,\langle8,6\rangle,\langle6,7\rangle,\langle5,10\rangle)$.

Next we consider the **modification** of the example **using cycle operations**. Fig. 5.6.6 shows the expanded cycle structure C^+ belonging to the cycle C inside the frame in Fig. 5.6.1.

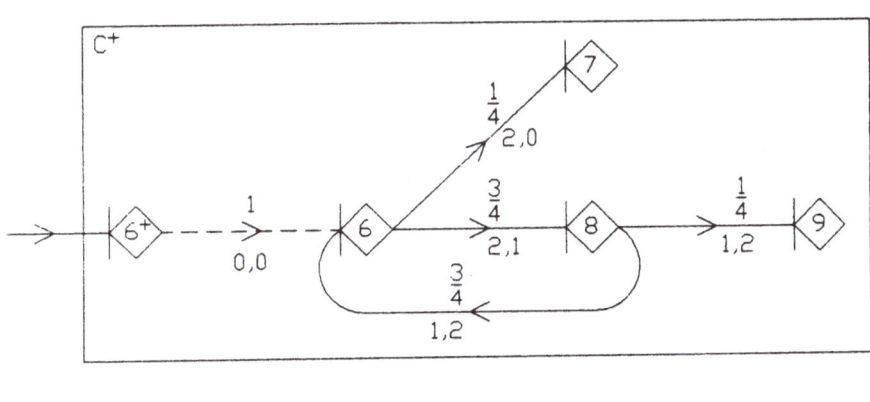

Figure 5.6.6

The progression through cycle C or respectively expanded cycle structure C^+ corresponds to a cycle operation o_6 with beginning event 6^+. The activation numbers z_{6+i} and transition probabilities q_{6+i} for all nodes i of C^+ different from the source 6^+ are listed in Table 5.6.2.

i	z_{6+i}	q_{6+i}
6	$16/7$	1
7	$4/7$	$4/7$
8	$12/7$	$3/4$
9	$3/7$	$3/7$

Table 5.6.2

By (5.5.4) and (5.5.5) we obtain for the expected duration τ_k and (mean) weight v_k of cycle operation o_k

$$\tau_6 = z_{6+6}p_{67}d_{67} + z_{6+6}p_{68}d_{68} + z_{6+8}p_{86}d_{86} + z_{6+8}p_{89}d_{89} = \frac{44}{7} \approx 6.29$$

$$v_6 = q_{6+6}p_{67}w_{67} + q_{6+6}p_{68}w_{68} + q_{6+8}p_{86}w_{86} + q_{6+8}p_{89}w_{89} = \frac{9}{4} = 2.25$$

Fig. 5.6.7 shows the acyclic EOR network resulting from the original network of Fig. 5.6.1 by expanding C to C^+ and replacing C^+ with the complete bipartite STEOR network with source set $R^+=\{6^+\}$ and sink set $S^+=\{7,9\}$.

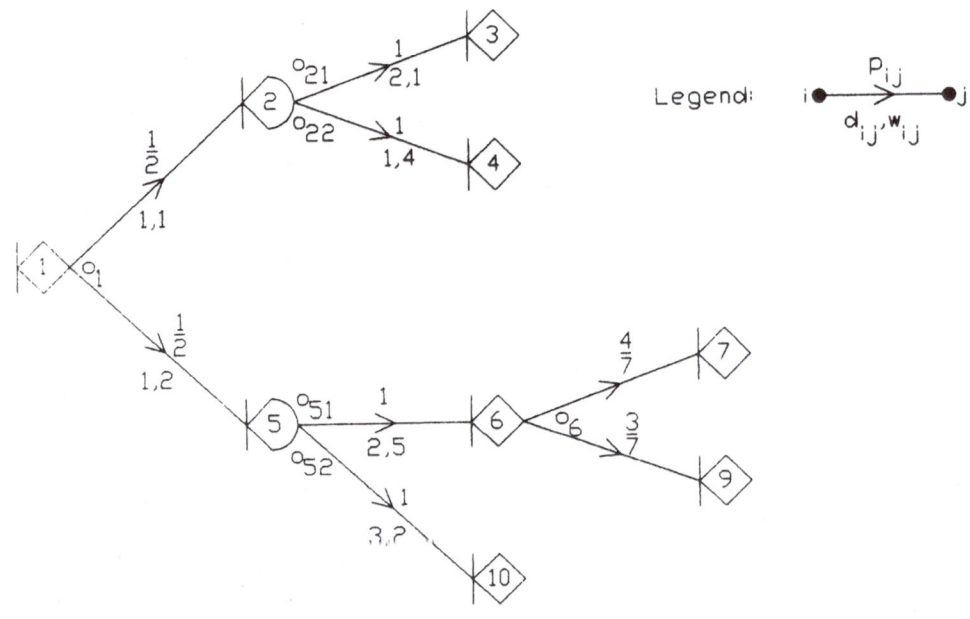

Figure 5.6.7

In comparison with the original problem, we have $\rho_6=2.79$ for cycle operation o_6 and there is no operation o_8. Iterations 1, 2 and 3 of algorithm (5.5.1) are as before and we again obtain $Q=(o_{22},o_{51},o_1)$. In iterations 4, 5 and 6 the operations o_5, o_{21} and o_6, respectively, are inserted at the rear of the current Q. This gives the optimal list schedule $Q=(o_{22},o_{51},o_1,o_{52},o_{21},o_6)$ and the optimal sequence of activity executions $(<1,5>,<5,6>,<5,10>,<6,8>,<8,6>,<6,7>)$ which is different from the activity sequence obtained for the original problem. Note that, for the project realization indicated by darker arrows in Fig. 5.6.1, the performance of cycle operation o_6 consists of the execution of activities $<6,8>,<8,6>$ and $<6,7>$ in that order.

5.7 Minimizing the Maximum Expected Lateness in EOR Networks

In this section we again follow closely Bücker and Neumann (1989) and Bücker et al. (1990). We present a polynomial algorithm, which computes an optimal precedence schedule for problem $1\,|\,\text{acyclEOR},D\text{-}G\,|\,\text{max}\mathbb{E}(L)$ (for the concept of a precedence schedule see (5.4.1) and for the objective function max $\mathbb{E}(L)$ compare (5.3.7) to (5.3.9)). The algorithm exploits Lawler's rule for solving the deterministic problem $1\,|\,\text{prec}\,|\,f_{\text{max}}$ (cf. (5.1.4)). In contrast to scheduling problem $1\,|\,\text{EOR},D\text{-}G\,|\,\mathbb{E}(\Sigma wC)$, the list schedule constructed for problem $1\,|\,\text{acyclEOR},D\text{-}G\,|\,\text{max}\mathbb{E}(L)$ does not necessarily represent an optimal solution (optimal scheduling policy) to the latter problem, which is NP–hard if we do not restrict ourselves to precedence schedules (cf. section 5.3). Whether the algorithm presented can be applied to EOR networks containing cycles, more general GERT networks, or more general objective functions will be discussed at the end of this section.

The expected lateness of operation o when precedence schedule Q is applied is denoted by $\mathbb{E}[L_o(Q)]$. Then $Q^*\in Q$ is an **optimal precedence schedule** if

(5.7.1) $\displaystyle \max_{o\in O} \mathbb{E}[L_o(Q^*)] = \min_{Q\in Q}\ \max_{o\in O} \mathbb{E}[L_o(Q)]$

where Q is the set of all precedence schedules of the acyclic admissible EOR network in question.

For what follows, so-called final operations turn out to be important. An operation $o \in O' \subseteq O$ is said to be a **final operation** of O' if there is no successor of o in O'. The sets of all final operations of O and O' are denoted by O_f and O'_f, respectively. For example, for the EOR network shown in Fig. 5.6.7, we have $O_f = \{o_{21}, o_{22}, o_{52}, o_6\}$. The last element of a precedence schedule is always a final operation. Since the following algorithm constructs an optimal precedence schedule backwards operation by operation, we need only consider final operations in that algorithm. For $o \in O_f$, the set of all precedence schedules whose last element is o is denoted by Q_o.

The algorithm that finds an optimal precedence schedule is based on a theorem which needs the following two lemmas.

(5.7.2)   Lemma.
Let $Q = (A, B, o, C)$, where A, B, and C are subsequences of a list schedule Q and o is an operation, and let $Q' = (A, o, B, C)$. Then $\mathbb{E}[L_o(Q')] \leq \mathbb{E}[L_o(Q)]$. Similarly, for $Q = (A, q, B, o, C)$ and $Q' = (A, B, o, q, C)$, where q is an operation, it holds that $\mathbb{E}[L_o(Q')] \leq \mathbb{E}[L_o(Q)]$.

Lemma (5.7.2) says that if an operation o is moved up a sequence of operations or if an operation q "before" o is moved down so that it occupies a position behind o without changing the order of the other operations, then the expected lateness of operation o gets smaller or remains the same. The proof of lemma (5.7.2) is trivial.

(5.7.3)   Lemma.
Let $o \in O_f$ and $Q, Q' \in Q_o$. Then $\mathbb{E}[L_o(Q)] = \mathbb{E}[L_o(Q')]$.

Lemma (5.7.3) says that all precedence schedules whose last element is the final operation o provide the same expected lateness of o. Lemma (5.7.3) is obvious because in both precedence schedules Q and Q' and for each project realization, operation o is performed at the same time.

(5.7.4)   **Theorem.**

Let $o^+ \in \mathcal{O}_f$ such that

(5.7.5) $$\mathbb{E}[L_{o^+}(\mathbb{Q}_{o^+})] = \min_{o \in \mathcal{O}_f} \mathbb{E}[L_o(\mathbb{Q}_o)]$$

where $\mathbb{Q}_o \in \mathcal{Q}_o$ for $o \in \mathcal{O}_f$. Then there is an optimal precedence schedule in $\mathcal{Q}_{o^+}$.

Theorem (5.7.4) states that for each final operation o^+ whose expected lateness is minimum, there is an optimal precedence schedule with last element o^+. Note that due to lemma (5.7.3), $\mathbb{Q}_{o^+} \in \mathcal{Q}_{o^+}$ and $\mathbb{Q}_o \in \mathcal{Q}_o$ can be chosen arbitrarily.

Proof.

Let $\mathbb{Q}_o = (A, o^+, B, o)$ where $o^+ \in \mathcal{O}_f$, $o \in \mathcal{O}_f \setminus \{o^+\}$ and o^+ satisfies (5.7.5). Moreover, let $\mathbb{Q}_{o^+} = (A, B, o, o^+)$. By lemma (5.7.2) we have

$$\mathbb{E}[L_o(\mathbb{Q}_{o^+})] \leq \mathbb{E}[L_o(\mathbb{Q}_o)]$$

and for each operation q from A or B

$$\mathbb{E}[L_q(\mathbb{Q}_{o^+})] \leq \mathbb{E}[L_q(\mathbb{Q}_o)]$$

Because of (5.7.5) it holds that

$$\mathbb{E}[L_{o^+}(\mathbb{Q}_{o^+})] \leq \mathbb{E}[L_o(\mathbb{Q}_o)]$$

So, for each $p \in \mathcal{O}$ there is an $r \in \mathcal{O}$ such that

$$\mathbb{E}[L_p(\mathbb{Q}_{o^+})] \leq \mathbb{E}[L_r(\mathbb{Q}_o)]$$

and thus

(5.7.6) $$\max_{p \in \mathcal{O}} \mathbb{E}[L_p(\mathbb{Q}_{o^+})] \leq \max_{p \in \mathcal{O}} \mathbb{E}[L_p(\mathbb{Q}_o)]$$

Inequality (5.7.6) says that $\mathbb{Q}_{o^+}$ is "better" than $\mathbb{Q}_o$ (compare (5.7.1)). Hence, there is an optimal precedence schedule whose last element is o⁺.

∎

Theorem (5.7.4) can be exploited as follows for constructing an optimal precedence schedule backwards: Let $\mathcal{O}' \subseteq \mathcal{O}$ be the set of unscheduled operations, and let $o^+ \in \mathcal{O}'_f$ such that (5.7.5) holds with $\mathcal{O}'_f$ instead of $\mathcal{O}_f$. Then o⁺ is to occupy the last position among the operations from $\mathcal{O}'$. This provides the following algorithm where the current precedence schedule $\mathbb{Q}$ is to be implemented as a linear list.

(5.7.7)　　**Algorithm.**

Step 1.　　Set $\mathbb{Q}:=\emptyset$ and $\mu:=m$

Step 2.　　Find $\mathcal{O}_f \subseteq \mathcal{O}$ and $o^+ \in \mathcal{O}_f$ such that $\mathbb{E}[L_{o^+}(\mathbb{Q}_{o^+})] = \min\limits_{o \in \mathcal{O}_f} \mathbb{E}[L_o(\mathbb{Q}_o)]$

　　　　　　where for $o \in \mathcal{O}_f$, $\mathbb{Q}_o \in \mathcal{Q}_o$ can be chosen arbitrarily

Step 3.　　Insert o⁺ at the top of $\mathbb{Q}$, delete o⁺ from $\mathcal{O}$ and set $\mu:=\mu-1$

　　　　　　If $\mu=0$, then terminate ($\mathbb{Q}$ is an optimal precedence schedule);

　　　　　　otherwise go to step 2　　　　　　　　　　　　　　　　　　　∎

In algorithm (5.7.7), m is again the number of operations of the underlying EOR network. Note that if o⁺ is deleted from $\mathcal{O}$ in step 3, the arcs from o⁺ have to be removed from the current EOR network and there generally results a different set $\mathcal{O}_f$ of final operations.

It remains to be shown how to compute the expected lateness $\mathbb{E}[L_o(\mathbb{Q}_o)]$ for $o \in \mathcal{O}_f$ where $\mathbb{Q}_o \in \mathcal{Q}_o$. Let q_i be the probability of occurrence of the beginning event i of operation o, let t_o be the expected start time of operation o, and let τ_o and δ_o be the expected duration and expected due date, respectively, of operation o. Then

(5.7.8)　　　　$\mathbb{E}[L_o(\mathbb{Q}_o)] = q_i(t_o+\tau_o-\delta_o)$

where

(5.7.9)
$$\begin{cases} \tau_o = \sum\limits_{<i,j> \in o} p_{ij}d_{ij} \\ \delta_o = \sum\limits_{<i,j> \in o} p_{ij}\delta_{ij} \end{cases}$$

and d_{ij} is again the expected duration of activity $<i,j>$.

Since q_i equals the expected number of activations z_i of node i, it holds that

$$q_1 = 1$$

(5.7.10) $$q_i = p_{1i} + \sum_{k=2}^{n} p_{ki} q_k \quad (i=2,\ldots,n)$$

where $V=\{1,\ldots,n\}$ is again the node set of the underlying acyclic EOR network N, node 1 is the source, and $p_{ki}:=0$ if there is no arc $<k,i>$ (compare (3.5.4)). If the nodes of the network N are ordered topologically (which can be done in $0(|E|)$ time), the system of linear equations (5.7.10) can easily be solved by "back–substitution", which takes $0(|V|^2)$ time (cf. (3.5.18)).

To compute t_o for $o \in O_f$, we consider an appropriate subnetwork N_o of the EOR network N in question, where N_o contains only arcs from operations that belong to the current set of operations O and are different from o. Then t_o is the expected time of carrying out all activities from N_o on a single machine. To define N_o precisely let $V(O)$ be the set of all beginning and terminal events of operations from O. Moreover, let $Z(i)$ be the set of the nodes reachable from any successor k of a deterministic node from $\mathcal{R}(i)$ where i is again the beginning event of operation o, $k \notin \mathcal{R}(i)$ and k is no terminal event of o. Then N_o is induced by the node set $V_o := V(O) \cap (\mathcal{R}(i) \cup Z(i))$. The reason for including the nodes from set $Z(i)$ in V_o is that, because o is to be a final operation, all operations different from o whose beginning events are reachable from a deterministic node from $\mathcal{R}(i)$ have to be performed before operation o can be started.

An example is illustrated in Fig. 5.7.1, which shows a subnetwork induced by node set $V(O)$. A stochastic operation o with beginning event i is indicated by dashed–line arrows and the corresponding subnetwork N_o by solid–line arrows. It holds that $Z(i)=\{1,2,3\}$. A deterministic operation o' with beginning event i' is given by the bold dashed–line arrow. The corresponding subnetwork $N_{o'}$ contains the bold solid–line arrows and we have $Z(i')=\{4,5,6,7\}$. The dotted–line arrows belong neither to N_o nor to $N_{o'}$.

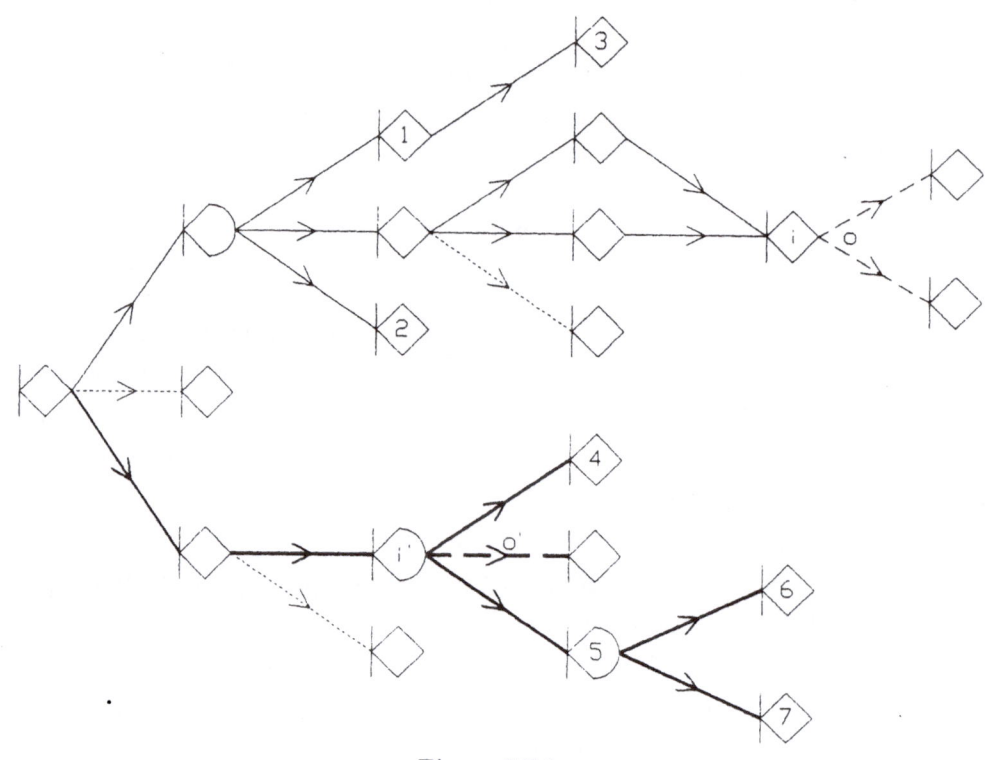

Figure 5.7.1

Let $\mathcal{S}_0(j):=\mathcal{S}(j)\cap V_0$, let E_0 be the arc set of N_0, and let

$$p_{jk}^0 := \begin{cases} \dfrac{p_{jk}}{\displaystyle\sum_{l\in\mathcal{S}_0(j)}p_{jl}} & \text{if } 0 < \displaystyle\sum_{l\in\mathcal{S}_0(j)}p_{jl}<1 \\[2em] p_{jk}, & \text{otherwise} \end{cases}$$

be the execution probability of activity $<j,k>$ and q_j^0 be the activation probability of node j in N_0. Then

$$(5.7.11) \qquad t_0 = \sum_{<j,k>\in E_0} q_j^0 p_{jk}^0 d_{jk}$$

The probabilities q_j^o $(j \in V_0)$ can be found by solving a system of linear equations of type (5.7.10), which takes $0(|V_0|^2)$ time. Since, for each operation, the calculations to be performed in algorithm (5.7.7) run in $0(|V|^2)$ time and the number of operations is $m=0(|E|)$, the time complexity of algorithm (5.7.7) is $0(|E||V|^2)$.

Since Lawler's rule solves the deterministic problem $1|prec|f_{max}$, the question arises whether algorithm (5.7.7) for problem $1|acyclEOR,D{\sim}G|maxE(L)$ can also be applied to scheduling problems with more general GERT precedence constraints and more general objective functions.

If the admissible acyclic GERT network in question contains a node k with IOR entrance, then some activities leading into node k may be carried out (and thus some operations "before" k may be performed) after the activation of k. Hence, the sequence of operations performed during some realization of an acyclic GERT network containing nodes with IOR entrance is not necessarily a subsequence of a precedence schedule. Therefore, the restriction to precedence schedules, which is basic to algorithm (5.7.7), does not make much sense if IOR nodes are present. Moreover, as already mentioned in section 5.3, problem $1|acyclGERT,D=1|maxE(L)$ is NP–hard even if we restrict ourselves to precedence schedules, which is caused by the presence of IOR nodes (cf. Bücker (1990)). If the acyclic GERT network possesses only nodes with AND entrance in addition to nodes with EOR entrance, algorithm (5.7.7) can be applied in principle. However, the computation of the quantities q_i and t_0 needed for determining $E[L_0(Q_0)]$ is much more complicated than in the case of an EOR network and cannot be done in polynomial time (cf. section 2.4 and Neumann and Steinhardt (1979), chapter 4).

Now suppose that the objective function $\max_{o \in O} E[L_0(Q)]$ is replaced by the more general function $\max_{o \in O} E[f_0(Q)]$ where f_0 is a nondecreasing cost function as specified in section 5.3 depending on the completion time of operation o. In this case the computation of the expected cost $E[f_0(Q)]$ of operation o given the precedence schedule Q requires the computation of the distribution function of the completion time of operation o which in general cannot be done in polynomial time. Again, algorithm (5.7.7) can be used in principle but it is no longer polynomial.

Finally we briefly discuss problem $1|\text{EOR},\text{D-G}|\max\mathbb{E}(L)$ where the admissible EOR
network N under consideration may contain cycles (and thus objective function (5.3.8)
instead of (5.3.9) is to be minimized). Algorithm (5.7.7) cannot be applied directly to N
because in each iteration, the algorithm selects a final operation of some subnetwork of
N, but an operation whose beginning event belongs to a cycle does not represent any
such final operation. Now assume that, as in the modification of problem
$1|\text{EOR},\text{D-G}|\mathbb{E}(\Sigma wC)$ from section 5.5, a progression through a cycle structure C of N when
C is entered at a specific entrance node is considered one cycle operation. Let o be any
operation with beginning event i, let l_o be the maximum expected lateness of operation
o if o has been begun at time 0, and let q_i, t_o, τ_o, and δ_o be as above. Then
$\mathbb{E}[L_o(\mathbb{Q}_o)]=q_i(t_o+l_o)$ where $l_o=\tau_o-\delta_o$ if o is no cycle operation (compare (5.7.8) and
(5.7.9)). For a cycle operation o from a cycle structure C, q_i can be found by deleting
all arcs emanating from node i and solving a system of linear equations, and t_o is
computed by (5.7.11) with z_j^o instead of q_j^o. The expected number of activations z_j^o of
node j in network N_o can again be found by solving a system of linear equations. To
compute l_o accurately, all possible realizations of C have to be enumerated (which
cannot be done in polynomial time) because any activity execution of C may be that one

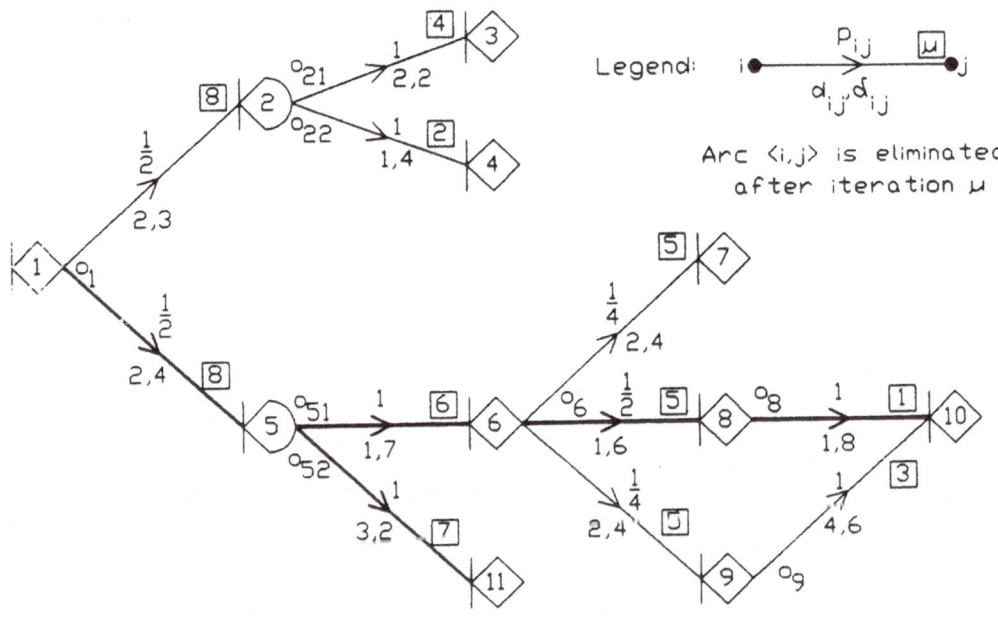

Figure 5.7.2

which provides the maximum lateness of the performances of operation o. In Bücker (1990) and Bücker and Neumann (1989) a polynomial algorithm is offered which produces an approximation for l_o. In that algorithm, cycle operation o is constructed successively by repeatedly linking "simple" stochastic operations, and the duration of each activity execution is supposed to be deterministic equal to the expected duration.

As an example of problem $1|\text{acyclEOR},D\text{-}G|\max\mathbb{E}(L)$ we consider the EOR network shown in Fig. 5.7.2. We want to compute an optimal precedence schedule with the aid of algorithm (5.7.7) and the corresponding sequence of activities carried out during the project realization indicated by darker arrows in Fig. 5.7.2. The algorithm starts with $Q=\emptyset$ and $O=\{o_1,o_{21},o_{22},o_{51},o_{52},o_6,o_8,o_9\}$.

At the beginning, the activation probabilities q_i for all nodes i aside from the sinks are computed successively. We obtain the values shown in Table 5.7.1.

i	1	2	5	6	8	9
q_i	1	$\frac{1}{2}$	$\frac{1}{2}$	$\frac{1}{2}$	$\frac{1}{4}$	$\frac{1}{8}$

Table 5.7.1

Iteration 1. The set of final operations is $O_f=\{o_{21},o_{22},o_{52},o_8,o_9\}$. First we have to compute

$$\Lambda_o := \mathbb{E}[L_o(Q_o)] = q_i(t_o+\tau_o-\delta_o)$$

(cf. (5.7.8)) for all $o\in O_f$ and determine $o^+\in O_f$ such that $\Lambda_{o^+}=\min_{o\in O_f} \Lambda_o$. For $o=o_{21}$ we get

$$q_i = q_2 = \frac{1}{2}, \ \tau_{o_{21}} = 2, \ \delta_{o_{21}} = 2$$

(compare (5.7.9)). To find $t_{o_{21}}$ we have to construct the network $N_{o_{21}}$ which is shown in Fig. 5.7.3. Then (5.7.11) gives $t_{o_{21}}=3$ and thus $\Lambda_{o_{21}} = \frac{3}{2}$. Analogously, we obtain

$$q_2 = \frac{1}{2}, \ \tau_{0_{22}} = 1, \ \delta_{0_{22}} = 4, \ t_{0_{22}} = 4, \ \Lambda_{0_{22}} = \frac{1}{2}$$

$$q_5 = \frac{1}{2}, \ \tau_{0_{52}} = 3, \ \delta_{0_{52}} = 2, \ t_{0_{52}} = 6, \ \Lambda_{0_{52}} = \frac{7}{2}$$

$$q_8 = \frac{1}{4}, \ \tau_{0_8} = 1, \ \delta_{0_8} = 8, \ t_{0_8} = 7, \ \Lambda_{0_8} = 0$$

$$q_9 = \frac{1}{8}, \ \tau_{0_9} = 4, \ \delta_{0_9} = 6, \ t_{0_9} = 8, \ \Lambda_{0_9} = \frac{3}{4}$$

where the networks $N_{0_{22}}$, $N_{0_{52}}$, N_{0_8}, and N_{0_9} are shown in Figs. 5.7.4 to 5.7.7.

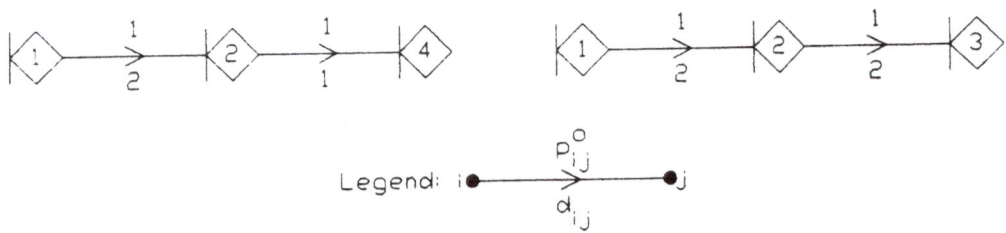

Figure 5.7.3. $N_{0_{21}}$ Figure 5.7.4. $N_{0_{22}}$

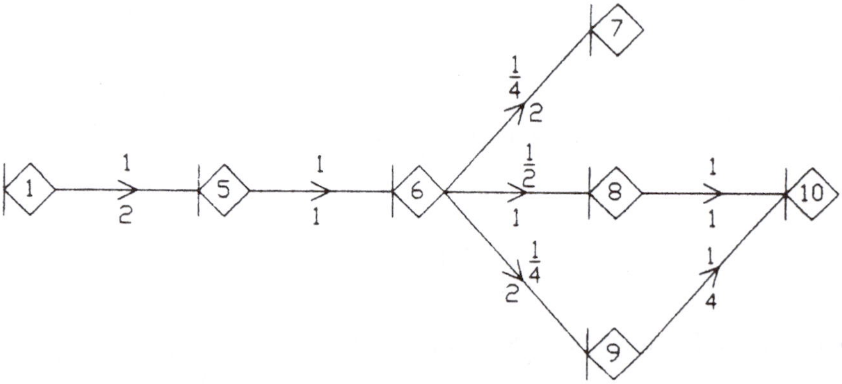

Figure 5.7.5. $N_{0_{52}}$

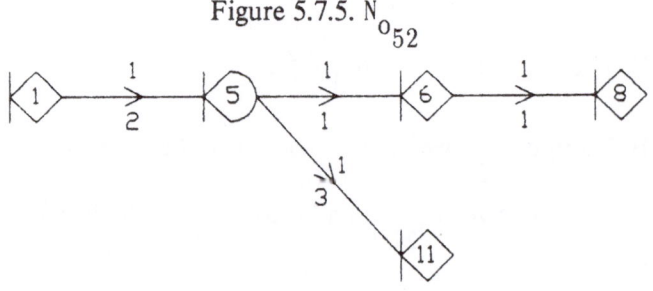

Figure 5.7.6. N_{0_8}

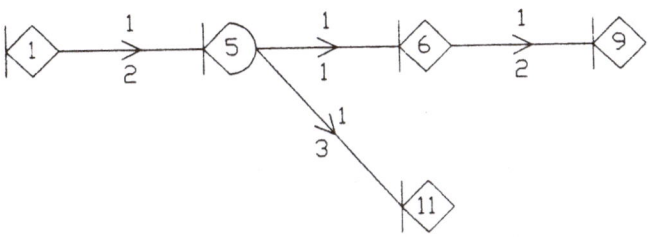

Figure 5.7.7. N_{o_9}

Since Λ_{o_8} is minimum, we have $o^+=o_8$, $\mathbb{Q}=(o_8)$ and $\mathcal{O}=\{o_1,o_{21},o_{22},o_{51},o_{52},o_6,o_9\}$. Moreover, arc $\langle 8,10\rangle$ has to be removed from the EOR network of Fig. 5.7.2.

Iteration 2. The new set of final operations is $\mathcal{O}_f=\{o_{21},o_{22},o_{52},o_9\}$. For the final operations $o=o_{21},o_{22},o_9$, the removed arc $\langle 8,10\rangle$ does not belong to N_o, and thus N_o and Λ_o are as before. For final operation o_{52}, network $N_{o_{52}}$ is shown in Fig. 5.7.8, and we obtain $t_{o_{52}}=\frac{11}{2}$ and $\Lambda_{o_{52}}=\frac{13}{4}$. $\Lambda_{o_{22}}$ is minimum and so $o^+=o_{22}$, $\mathbb{Q}=(o_{22},o_8)$ and $\mathcal{O}=\{o_1,o_{21},o_{51},o_{52},o_6,o_9\}$. Arc $\langle 2,4\rangle$ and node 4 are removed from the EOR network.

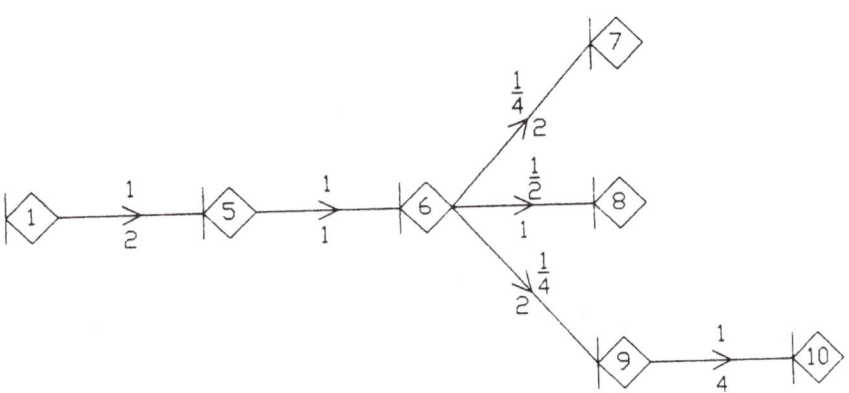

Figure 5.7.8. $N_{o_{52}}$

Iteration 3. We have $\mathcal{O}_f=\{o_{21},o_{52},o_9\}$. As before, $\Lambda_{o_{52}}=\frac{13}{4}$ and $\Lambda_{o_9}=\frac{3}{4}$. The new network $N_{o_{21}}$ is shown in Fig. 5.7.9, and we obtain $t_{o_{21}}=2$ and $\Lambda_{o_{21}}=1$. Moreover,

$o^+=o_9$, $\bar{q}=(o_9,o_{22},o_8)$, $\bar{o}=\{o_1,o_{21},o_{51},o_{52},o_6\}$, and arc $<9,10>$ and node 10 are eliminated.

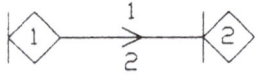

Figure 5.7.9. $N_{o_{21}}$

Iteration 4. We have $\bar{o}_f=\{o_{21},o_{52},o_6\}$. As before, $\Lambda_{o_{21}}=1$. For final operation o_{52}, network $N_{o_{52}}$ is shown in Fig. 5.7.10, and we obtain $t_{o_{52}}=\frac{9}{2}$ and $\Lambda_{o_{52}}=\frac{11}{4}$. For operation o_6, we have

$$q_6=\frac{1}{2},\ \tau_{o_6}=\frac{3}{2},\ \delta_{o_6}=5,\ t_{o_6}=6,\ \Lambda_{o_6}=\frac{5}{4}$$

where network N_{o_6} is given in Fig. 5.7.11. Hence, $o^+=o_{21}$, $\bar{q}=(o_{21},o_9,o_{22},o_8)$, $\bar{o}=\{o_1,o_{51},o_{52},o_6\}$, and arc $<2,3>$ and node 3 are removed.

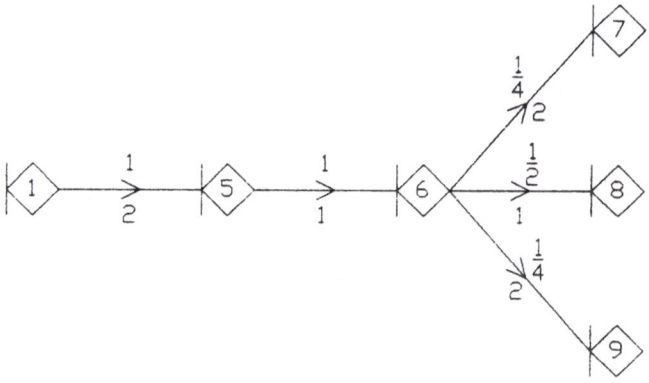

Figure 5.7.10. $N_{o_{52}}$

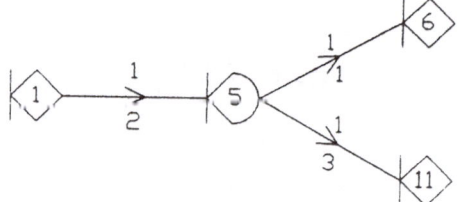

Figure 5.7.11. N_{o_6}

Iteration 5. We have $O_f=\{o_{52},o_6\}$. $\Lambda_{o_{52}}$ and Λ_{o_6} are as before. We get $o^+=o_6$, $Q=(o_6,o_{21},o_9,o_{22},o_8)$, $O=\{o_1,o_{51},o_{52}\}$, and arcs $<6,7>$, $<6,8>$ and $<6,9>$ together with their final nodes are eliminated.

Iteration 6. We obtain $O_f=\{o_{51},o_{52}\}$, $\Lambda_{o_{51}} = -\frac{1}{2}$ and $\Lambda_{o_{52}} = 2$. The networks $N_{o_{51}}$ and $N_{o_{52}}$ are shown in Figs. 5.7.12 and 5.7.13, respectively. Moreover, $o^+=o_{51}$, $Q=(o_{51},o_6,o_{21},o_9,o_{22},o_8)$, $O=\{o_1,o_{52}\}$, and arc $<5,6>$ and node 6 are removed.

Figure 5.7.12. $N_{o_{51}}$ Figure 5.7.13. $N_{o_{52}}$

In *iteration 7* we have $O_f=\{o_{52}\}$ and in *iteration 8* $O_f=\{o_1\}$. Thus, $Q=(o_1,o_{52},o_{51},o_6,o_{21},o_9,o_{22},o_8)$ is an optimal precedence schedule, and the corresponding sequence of activities carried out during the project realization indicated by darker arrows in Fig. 5.7.2 is $(<1,5>,<5,11>,<5,6>,<6,8>,<8,10>)$.

5.8   Essential Histories and Scheduling Policies for Min–Sum Problems in General GERT Networks

In the preceding sections we have presented algorithms that provide optimal list schedules or respectively precedence schedules for special scheduling problems with EOR precedence constraints. For the min–sum problem $1|EOR,D\text{-}G|E(\Sigma wC)$, an optimal list schedule also represents an optimal solution. For general GERT precedence constraints or more general objective functions of the type $E(\Sigma f)$, the latter fact does not hold any longer and we have to consider general scheduling policies.

In what follows, we first make precise the concept of a scheduling policy. Then we assign a Markov renewal process to problem $1|GERT,D\text{-}G|E(\Sigma f)$, which will be exploited in section 5.10 for computing an optimal scheduling policy. We will only deal with min–sum scheduling problems from now on because min–max problems are even less tractable (compare Bücker (1990)).

As already stated in (5.2.2), loosely speaking, a **scheduling policy** specifies a performable operation (the operation to be performed next) given the project evolution up to the present. The project evolution up to the present, say by time t, includes the project state at time t and the past history up to time t (for the latter two concepts compare (1.3.5)). If a scheduling policy is independent of the past history, it is called a **Markov scheduling policy**. For example, a list schedule, which assigns a fixed priority to each operation, represents a special Markov scheduling policy (cf. (5.4.1)).

To give the complete project evolution up to the present is very cumbersome. Thus, we are interested in finding out which part of the project evolution is sufficient for uniquely specifying scheduling policies for problem $1|\text{GERT},D\text{-}G|\mathbb{E}(\Sigma f)$ and which will be called the **essential history**. If we then enlarge the underlying state space such that the essential histories constitute the (extended) states, we can assign a Markov renewal process to each scheduling policy. This permits us to use a dynamic programming approach for computing an optimal scheduling policy.

(5.8.1) The essential history should contain the following information:
(a)   All EOR nodes activated last (a node i is said to be **activated last** if i has been activated, say, for the βth time but no activity with initial node i has been carried out after the βth activation of i)
(b)   All IOR nodes activated (it is not sufficient to include only the IOR nodes activated last because if an activity is carried out whose final node is an IOR node i already activated earlier, node i must not be activated again and thus no operation with beginning event i will again become performable)
(c)   All activities carried out last whose final node is an AND node (an activity $\langle i,j\rangle$ is said to be **carried out last** if $\langle i,j\rangle$ has been executed, say, for the αth time but no activity with initial node j has been carried after the αth execution of $\langle i,j\rangle$)

(d)   All activities carried out whose initial node is deterministic (if some of the
       activities with one and the same deterministic initial node have been carried out,
       the remaining ones represent performable operations) [20].

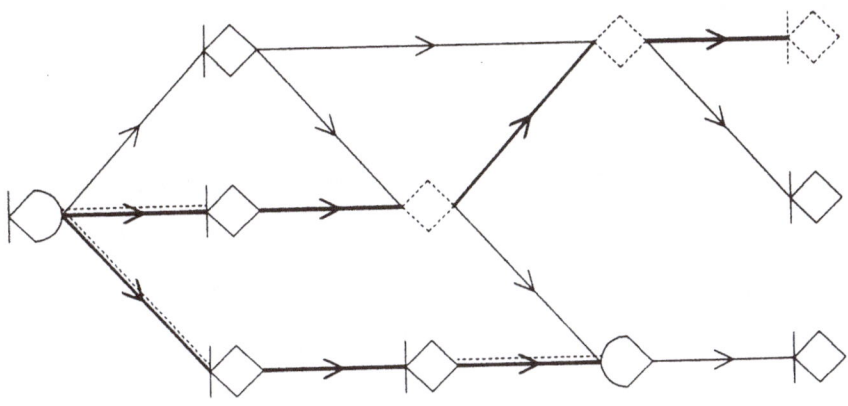

Figure 5.8.1

As we see from (5.8.1b) and (5.8.1d), the essential history includes some past history.
Fig. 5.8.1 shows a GERT network (without arc weights), where the activities carried
out are indicated by darker arrows. The arcs and nodes to be included in the essential
history according to (5.8.1a) to (5.8.1d) are dotted. The construction of the "dotted"
set of nodes and arcs and its updating as the project evolution proceeds is in general
very cumbersome. Instead, Rubach (1984) has considered the set of all activities outside
cycles that have been carried out so far and of all nodes activated last inside cycles. The
latter set, which corresponds to the concept of a "feasible set of jobs" for the
deterministic problem $1|\text{prec}|\Sigma f_\nu$ introduced by Schrage and Baker (1978) and by

Lawler (1979), is larger than the "dotted" set but can be handled more easily. Before
we deal with that set in more detail, we introduce some notation.

(5.8.2) Recall that the activities of the project have to be carried out one after another
because each activity must pass through the single machine available. Let $<\kappa>$ denote
the κth activity execution in turn ($\kappa=1,2,\ldots$). Note that each activity outside any
cycle which is carried out during the project execution in question corresponds to only

---

[20]   If all activities with one and the same deterministic initial node have been carried
        out and for each of them at least one "following" activity has been executed (that
        is, they do not represent activities carried out last), these activities may be
        removed from the essential history.

one index κ. An activity within a cycle, however, which may be carried out several times, can be associated with several indices. Let D_κ be the duration and C_κ be the completion time of activity execution $<\kappa>$. Since there is no idle time between consecutive activity executions, we have

$$C_\kappa = C_{\kappa-1} + D_\kappa \quad (\kappa=1,2,\ldots)$$

where $C_0:=0$. Moreover, let M be the number of activity executions during a single project execution. Note that D_κ, C_κ $(\kappa=1,\ldots,M)$, and M are random variables defined on the sample space Ω, the set of all possible project realizations. C_M equals the project duration D. Furthermore, we have $P(M<\infty)=1$ (cf. theorem (1.3.9)).

(5.8.3) We describe a sequence of activity executions $<1>,\ldots,<\kappa>$ during a project realization $\omega\in\Omega$ (and thus the essential history up to and including time $C_\kappa(\omega)$) by the following **feasible arc–node set** $\mathcal{X}_\kappa(\omega)$: $\mathcal{X}_\kappa(\omega)$ contains all activities outside cycles that have been carried out by time $C_\kappa(\omega)$ and all nodes activated last that belong to any cycle. For $\kappa=0$ we put $\mathcal{X}_0(\omega):=\emptyset$. The set of all feasible arc–node sets of the underlying GERT network is denoted by $\mathfrak{X}$:

$$\mathfrak{X} := \{\mathcal{X}_\kappa(\omega) \,|\, 0\leq\kappa\leq M(\omega),\ \omega\in\Omega\}$$

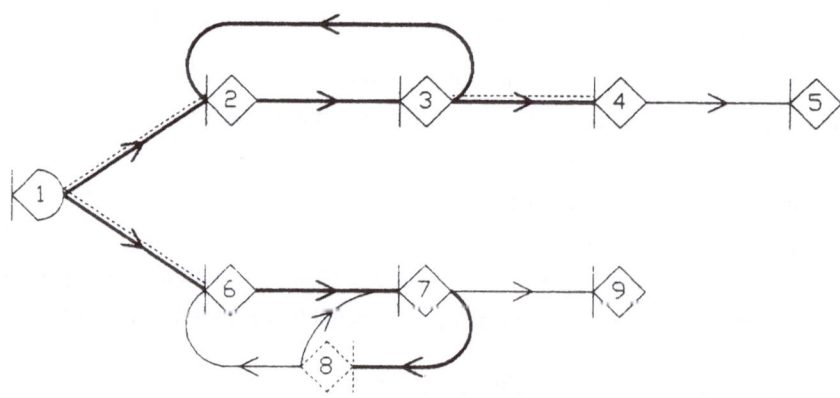

Figure 5.8.2

The elements of $\mathfrak{X}$ will be designated by x in what follows. Fig. 5.8.2 shows a GERT network (without arc weights), where a sequence of activity executions is again indicated by bold–line arrows. The feasible arc–node set associated with that sequence contains the dotted–line arcs <1,2>,<3,4> and <1,6> and the dotted node 8.

(5.8.4) Next, we make precise what is to be understood by a performable operation in the present more general context (compare (5.2.5)). We say that a feasible arc–node set x **activates** a node i when x contains *all* arcs leading into i if i is an AND node, *at least one* incoming arc if i is an IOR node, or *exactly one* incoming arc if i is an EOR node. Given $x \in \mathfrak{X}$ an operation $o \in \mathcal{O}$ with beginning event i is called **performable** if node i belongs to x (then i lies in a cycle) or if i does not belong to x, x activates node i, and x does not contain any arc from o. The set of all performable operations for given x is denoted by $\mathcal{O}(x)$. In Fig. 5.8.2 the operations {<4,5>} and {<8,6>,<8,7>} are performable.

(5.8.5) Given a feasible arc–node set x, the execution of an activity <i,j> from a performable operation $o \in \mathcal{O}(x)$ provides the following feasible arc–node set $y_{ij}(x)$ depending on the preceding set x and the activity <i,j> carried out:

$$
y_{ij}(x) := \begin{cases} (x \cup \{j\}) \setminus \{i\} & \text{if <i,j> belongs to a cycle structure} \\ x \cup \{<i,j>\} \cup \{j\} & \text{if <i,j> is an entrance arc of a cycle structure} \\ (x \cup \{<i,j>\}) \setminus \{i\} & \text{otherwise} \end{cases}
$$

$y_{ij}(x)$ is called the (immediate) **follower of x owing to activity** <i,j>. In the network of Fig. 5.8.2 let again

$$x = \{<1,2>,<3,4>,<1,6>,8\}$$

Then

$$y_{86}(x) = \{<1,2>,<3,4>,<1,6>,6\}$$

is the follower of x owing to activity <8,6>. The set of all followers of $x \in \mathfrak{X}$ is denoted by $\mathcal{F}(x)$:

$$\mathcal{F}(x) := \{y_{ij}(x) \mid <i,j>\in o, o\in \mathcal{O}(x)\}$$

For the network of Fig. 5.8.2 we have $\mathcal{F}(x)=\{y_{86}(x), y_{87}(x), y_{45}(x)\}$, where

$$y_{87}(x) = \{<1,2>,<3,4>,<1,6>,7\}$$
$$y_{45}(x) = \{<1,2>,<3,4>,<4,5>,<1,6>,8\}$$

A feasible arc–node set y is said to be a **follower in the broader sense** of $x\in\mathcal{X}$, in symbols $y\in\bar{\mathcal{F}}(x)$, if $y\in\mathcal{F}(x)$ or there exist $x^1,\ldots,x^r\in\mathcal{X}$ such that $x^1\in\mathcal{F}(x), y\in\mathcal{F}(x^r)$, and $x^{\rho+1}\in\mathcal{F}(x^\rho)$ for $\rho=1,\ldots,r-1$. In other words, $\bar{\mathcal{F}}(x)$ is the transitive closure of $\mathcal{F}(x)$.

(5.8.6) If the skipping scheduling problem $1|\text{GERT},\text{skip},D\text{-}G|\mathbb{E}(\Sigma f)$ has to be solved, then, given a feasible arc–node set x, operations from $\mathcal{O}(x)$ must not be performed any longer when no more sink can be activated. To do so we proceed as follows. Let $x\in\mathcal{X}$ and let $S(x)$ be the set of those sinks of the GERT network in question which are activated by x. Trivially, $S(y)\supseteq S(x)$ for all $y\in\bar{\mathcal{F}}(x)$. We then put $\mathcal{O}(x):=\emptyset$ if there is no $y\in\bar{\mathcal{F}}(x)$ such that $S(y)\supset S(x)$ (that is, if no $y\in\bar{\mathcal{F}}(x)$ activates at least one sink in addition to the sinks activated by x).

(5.8.7) Now we introduce the concept of an (extended) project state and make precise the concept of a scheduling policy. For $x\in\mathcal{X}$, a point in time $s\in\mathbb{R}_+$ is called a **completion time** of x if there exist an $\omega\in\Omega$ and a $\kappa\in\{0,1,\ldots,M(\omega)\}$ such that $x=\mathcal{X}_\kappa(\omega)$ and $s=C_\kappa(\omega)$.

If the activity durations are stochastic or cycles are present, there are in general several completion times for a given x. Thus, instead of x alone, a pair (x,s) where $x\in\mathcal{X}$ and s is any completion time of x is considered a **project state** in what follows. Then $\mathcal{X}\times\mathbb{R}_+$ is the **state space**. A mapping $\phi:\mathcal{X}\times\mathbb{R}_+\to\mathcal{O}$ that assigns a performable operation $\phi(x,s)\in\mathcal{O}(x)$ to each project state (x,s) is called a **scheduling policy**. Given a scheduling policy and a project realization, we can easily construct the corresponding sequence of activity executions.

(5.8.8) We consider the special case where the admissible GERT network under consideration is acyclic and the duration of each activity $<i,j>$ is deterministic, say, equal to d_{ij}. Then each feasible arc–node set x contains only arcs and its completion time s is unique and given by

$$(5.8.9) \qquad s = \sum_{\langle i,j \rangle \in x} d_{ij}$$

Hence, we may restrict ourselves to **stationary scheduling policies** $\phi:\mathfrak{X}\to\mathcal{O}$, that is, scheduling policies which are not explicitly dependent upon time. List schedules, which have turned out to be sufficient for solving the scheduling problem $1|\text{EOR},\text{D}\sim\text{G}|\mathbb{E}(\Sigma wC)$, can also be viewed as special stationary policies. Let $\mathbb{Q}$ be a list schedule and, for $x\in\mathfrak{X}$, let $\mathbb{Q}(x)$ be that subsequence of $\mathbb{Q}$ which contains exactly the elements of $\mathcal{O}(x)$. Then the function ϕ where $\phi(x)$ is the first element of $\mathbb{Q}(x)$ is a stationary scheduling policy. In Rubach (1984) is shown that for solving the scheduling problem $1|\text{GERT},\text{D}\sim\text{G}|\mathbb{E}(\Sigma wC)$, we can restrict ourselves to stationary scheduling policies as well.

(5.8.10) The selection of a scheduling policy ϕ induces a Markov renewal process $(\mathcal{X}_\kappa,C_\kappa)^\phi_{\kappa\in\mathbb{N}_0}$ with state space $\mathfrak{X}\times\mathbb{R}_+$ where we put

$$\left.\begin{array}{l} \mathcal{X}_\kappa := \mathcal{X}_M \\ C_\kappa := C_M \end{array}\right\} \text{ for } \kappa > M$$

in order to define $\mathcal{X}_\kappa$ and C_κ for all $\kappa\in\mathbb{N}_0$ (for Markov renewal processes we again refer to section 3.1). To specify the transition functions $\mathbb{Q}^\phi_{xy}$ $(x,y\in\mathfrak{X})$ of the Markov renewal process $(\mathcal{X}_\kappa,C_\kappa)^\phi_{\kappa\in\mathbb{N}_0}$, we note that to each pair (x,y) where $y=y_{ij}(x)$ is the follower of x owing to activity $\langle i,j \rangle$, there corresponds exactly one arc of the underlying GERT network N, namely the arc $\langle i,j \rangle$. Then for $s,t\geq0$ and $x,y\in\mathfrak{X}$

$$(5.8.11) \qquad \mathbb{Q}^\phi_{xy}(s,t) := P(\mathcal{X}_{\kappa+1}=y, C_{\kappa+1}\leq s+t \,|\, \mathcal{X}_\kappa=x, C_\kappa=s)^\phi$$

$$= \left\{ \begin{array}{ll} p_{ij}F_{ij}(t) & \text{if } y=y_{ij}(x) \text{ and } \langle i,j \rangle\in\phi(x,s) \\ 1 & \text{if } \mathcal{O}(x)=\emptyset \text{ and } x=y \\ 0 & \text{otherwise} \end{array} \right.$$

where $\begin{bmatrix} P_{ij} \\ F_{ij} \end{bmatrix}$ is the weight of arc $<i,j>$ in N. Since the transition function q_{xy}^{ϕ} depends

on the completion time s of x, the Markov renewal process $(X_{\kappa}, C_{\kappa})_{\kappa \in \mathbb{N}_0}^{\phi}$ is

nonstationary. For such a family of Markov renewal processes (with family parameter ϕ), we also use the term **Markov renewal decision process**, where the word "decision" indicates that we are looking for an **optimal scheduling policy** that minimizes the objective function $\mathbb{E}(\Sigma f)$. To solve the scheduling problem $1|GERT,D\text{-}G|\mathbb{E}(\Sigma f)$, a stochastic dynamic programming approach can then be employed, which will be presented in section 5.10. Some elements of dynamic programming which are needed in section 5.10 will be summarized in section 5.9.

(5.8.12) Since a Markov renewal process can be associated with each scheduling policy and Markov renewal processes are related to STEOR networks (compare section 3.2), it suggests itself to associate a STEOR network with the scheduling problem $1|GERT,D\text{-}G|\mathbb{E}(\Sigma f)$. This STEOR network, which will be considered in (6.2.8) in more detail, has node set $\mathfrak{X}$, arc set $\{<x,y>|x\in\mathfrak{X}, y\in\mathcal{F}(x)\}$, and weights of the arcs $<x,y>$ that depend on the completion time s of x and the scheduling policy ϕ selected.

5.9   Elements of Dynamic Programming

This section gives a brief review of dynamic programming necessary to understand the method of solving problem $1|GERT,D\text{-}G|\mathbb{E}(\Sigma f)$ presented in section 5.10. For further information on deterministic and stochastic dynamic programming we refer to the relevant literature such as Bertsekas (1976), Denardo (1982), Heyman and Sobel (1984), chapters 4 and 5, Hinderer (1970), Neumann (1977), chapter 1, and Whittle (1982, 1983).

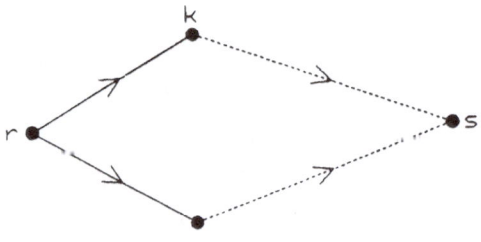

Figure 5.9.1

The basic idea of (deterministic) dynamic programming can be demonstrated best by looking for the shortest path in a network with weights $c_{ij} \in \mathbb{R}$ of the arcs $<i,j>$ where we assume that the network does not contain cycles of negative length. Let s be a given node of the network and let d_r be the length of a shortest path from any node r in the network to s where $s \in \mathcal{Z}(r)$. If k is the successor of r on such a shortest path (see Fig. 5.9.1), then the part of that path from k to s must also be a shortest path (otherwise the overall path from r to s would not be as short as possible), that is, we have

(5.9.1) $d_r = c_{rk} + d_k$

This is Bellman's so-called **principle of optimality**. Clearly, in (5.9.1) k has to be a node for which $c_{rk} + d_k$ is as small as possible. Thus, the shortest path lengths must satisfy the following equation

(5.9.2) $d_r = \min_{k \in \mathcal{S}(r)} (c_{rk} + d_k)$

which is referred to as **Bellman's equation**. Bellman's equation (5.9.2) also represents a sufficient condition for optimality if every cycle in the network has a positive length.

Now we want to present the basic features which characterize dynamic programming problems in a more general way. Consider a system which moves from stage to stage over time, where we are supposed to have an infinite planning horizon. Each stage has a set of states associated with it, which represent the various possible conditions in which the system might be at that stage. The set of all possible states of the system is again called the **state space** and is denoted by $\mathcal{X}$. At each stage, a decision u has to be made, which results in a transformation of the current state x into a state associated with the next stage. The latter state, which is also called a **successor** of state x and may depend on x and on the decision u, is designated by $y(x,u)$. The set of all possible decisions u may also depend on x and is denoted by $U(x)$.

Suppose a cost is incurred if decision u is made. This cost, which can depend on the decision u and current state x, is denoted by $h(x,u)$. Then we are interested in such a sequence of decisions (called on **optimal sequence of decisions**) that the total cost is minimized if we start at any state x.

Note that all quantities related to the transformation from the current stage into the next stage (set of decisions $U(x)$, successor $y(x,u)$ of x, and cost $h(x,u)$) depend only on the state x at the current stage and not on the previous behaviour of the system. Hence, an optimal sequence of decisions for the remaining stages is independent of the decisions made at previous stages. This is again the **principle of optimality**, which says that the underlying process is memoryless and which represents some kind of Markov property.

In the special problem of finding shortest paths in a network, the nodes of the network correspond to states, a decision consists of choosing a successor of the current node, and the minimum total cost corresponds to the shortest path length.

In the more general dynamic programming model, let $H^*(x)$ be the minimum total cost when starting at state x. Then we have in analogy to (5.9.2)

(5.9.3) $$H^*(x) = \min_{u \in U(x)} \{h(x,u) + H^*[y(x,u)]\} \quad (x \in \mathfrak{X})$$

Let $\phi^*(x)$ denote an optimal decision made in state x, that is, a minimizer of the expression within the braces in (5.9.3). Then the corresponding function ϕ^* defined on the state space $\mathfrak{X}$ is called an **optimal policy**. Given an optimal policy ϕ^* and a beginning state x_0, an optimal sequence of decisions $u_1, u_2, \ldots$ to be made at consecutive stages and the corresponding sequence of states $x_1, x_2, \ldots$ can be found as follows:

$$u_1 = \phi^*(x_0), \; x_1 = y(x_0, u_1)$$
$$u_2 = \phi^*(x_1), \; x_2 = y(x_1, u_2)$$
$$\ldots$$

Note that ϕ^* represents a stationary policy, that is, it does not depend on the number of the stage at which the decision is to be made. This results from the fact that the quantities $U(x)$, $y(x,u)$, and $h(x,u)$ are independent of the stage number.

For conditions on the sets $\mathfrak{X}$ and $U(x)$ and functions h and y which ensure first that Bellman's equation (5.9.3) has a unique solution $H^*(x)$ which represents the minimum total cost and second that there is a minimizing function ϕ^* which provides an optimal sequence of decisions, we refer to the literature mentioned above. Also, methods for computing an optimal policy can be found there.

(5.9.4) Thus far, we have considered a deterministic dynamic programming problem. Now suppose that the transition from the current state x to a successor y occurs according to a probability distribution, say, with a probability $p_{xy}(u)$, which depends on the decision u made. Assuming that the state space $\mathcal{X}$ is finite or countably infinite, the principle of optimality implies that, for any u, the probabilities $p_{xy}(u)$ $(x,y \in \mathcal{X})$ are the transition probabilities of a (homogeneous) Markov chain (cf. section 3.1). Let $\tilde{h}_{xy}(u)$ be the cost incurred when the system moves from state x to state y owing to decision u. Furthermore, let

$$h(x,u) := \sum_{y \in \mathcal{X}} p_{xy}(u) \tilde{h}_{xy}(u)$$

be the expected cost arising when a transition from state x to some other state occurs and decision u has been made, and let $H^*(x)$ be the minimum expected total cost when we start at state x. Then Bellman's equation (5.9.3) gets the form

$$(5.9.5) \qquad H^*(x) = \sum_{u \in U(x)} \{h(x,u) + \sum_{y \in \mathcal{X}} p_{xy}(u) H^*(y)\} \quad (x \in \mathcal{X})$$

For methods of solving Bellman's equation (5.9.5) and computing an optimal policy (in case there exist a solution to (5.9.5) and an optimal policy) and for more general stochastic dynamic programming problems we again refer to the literature cited above.

5.10 Determination of an Optimal Scheduling Policy for the General Min–Sum Problem

The problem of finding an optimal scheduling policy for the scheduling problem $1|GERT,D\text{-}G|\mathbb{E}(\Sigma f)$ represents a stochastic dynamic programming problem which is a little more complicated than that one considered in (5.9.4). In the former problem, we have a Markov renewal process $(X_\kappa, C_\kappa)^\phi_{\kappa \in \mathbb{N}_0}$ with state space $\mathcal{X} \times \mathbb{R}_+$ instead of a Markov chain where $\mathfrak{X}$ is the set of all feasible arc-node sets. The imbedded Markov chain $(\mathcal{X}_{\kappa})_{\kappa \in \mathbb{N}_0}^{\phi}$ has the transition probabilities

$$(5.10.1) \qquad p_{xy}^{\phi} := \begin{cases} p_{ij} & \text{if } y=y_{ij}(x) \text{ and } <i,j> \in \phi(x,s) \\ 1 & \text{if } \mathcal{O}(x)=\emptyset \text{ and } x=y \qquad (x,y \in \mathfrak{X}) \\ 0 & \text{otherwise} \end{cases}$$

(compare (5.8.11)). C_{κ} is a continuous random variable with range $\mathbb{R}_{+}$, and the conditional distribution function of $D_{\kappa}=C_{\kappa}-C_{\kappa-1}$ given that activity $<i,j>$ is carried out (representing the κth activity execution in turn) is F_{ij}. The transition from a state (x,s) to a successor $(y_{ij}(x),s+t)$ depends on the performable operation $o \in \mathcal{O}(x)$ selected (which plays the role of the decision to be made), the activity $<i,j>$ carried out in operation o, and the (realized) duration t of activity $<i,j>$.

Next, we explain the form of Bellman's equation for the scheduling problem $1|GERT,D-G|\mathbb{E}(\Sigma f)$ without giving a precise derivation. In chapter 6 we will see that this scheduling problem can be viewed as a special case of a general cost optimization problem for STEOR networks (cf. (6.2.8)) and we will there discuss in more detail how to solve that problem by means of dynamic programming techniques.

The expected cost of performing operation o if o begins at time s is

$$(5.10.2) \qquad h_o(s) = \sum_{<i,j> \in o} p_{ij} \int_{\mathbb{R}_+} f_{ij}(s+t) F_{ij}(dt)$$

The expected total cost of the "partial project" starting at state (x,s) if policy ϕ is used is denoted by $H_{\phi}(x,s)$ where

$$H_{\phi}(x,s) = 0 \quad \text{if} \quad \mathcal{O}(x) = \emptyset$$

In (6.3.2) we will see that $H_{\phi}(x,s)$ can be expressed in terms of $h_o(s)$ and the multi–step transition functions of the Markov renewal process $(\mathcal{X}_\kappa, C_\kappa)^\phi_{\kappa\in\mathbb{N}_0}$ corresponding to policy ϕ.

Let Φ be the set of all scheduling policies ϕ. We then seek to determine the **minimum total cost function** H^* given by

$$H^*(x,s) := \inf_{\phi\in\Phi} H_\phi(x,s) \quad \text{for all } (x,s)\in\mathcal{X}\times\mathbb{R}_+$$

and an **optimal scheduling policy** $\phi^*\in\Phi$ with $H_{\phi^*}=H^*$. Since $x=\emptyset$ and $s=0$ at the beginning of each project execution, $H^*(\emptyset,0)$ is the infimum expected total cost of the project.

Bellman's equation (5.9.5) now takes the form

(5.10.3) $\qquad H^*(x,s) = \begin{cases} \displaystyle\min_{o\in\mathcal{O}(x)} \{h_o(s)+ \sum_{<i,j>\in o} p_{ij}\int_{\mathbb{R}_+} H^*(y_{ij}(x),s+t)F_{ij}(dt)\} \\ \hspace{5cm} \text{for } \mathcal{O}(x)\neq\emptyset \\[2mm] 0 \quad \text{for } \mathcal{O}(x)=\emptyset \end{cases}$

$$((x,s)\in\mathcal{X}\times\mathbb{R}_+)$$

In (5.10.3) we may write "min" instead of "inf" because the sets $\mathcal{O}(x),x\in\mathcal{X}$, are finite. $\phi^*(x,s)$ represents a minimizer of the expression within the braces in (5.10.3) for given (x,s).

We consider the special case where the GERT network in question is acyclic and the duration of each activity $<i,j>$ is deterministic, say, equal to d_{ij} (compare (5.8.8)). Then each feasible arc–node set $x\in\mathcal{X}$ contains only arcs, and the follower $y_{ij}(x)$ of x owing to activity $<i,j>$ equals $x\cup\{<i,j>\}$. Moreover, the unique completion time s of any $x\in\mathcal{X}$ is given by (5.8.9), and (5.10.2) gets the simpler form

$$h_o(s) = \sum_{<i,j>\in o} p_{ij}f_{ij}(s+d_{ij})$$

Thus, Bellman's equation (5.10.3) reduces to

$$(5.10.4) \; H^*(x) = \begin{cases} \min_{o \in \mathcal{O}(x)} \; \sum_{<i,j> \in o} p_{ij} [f_{ij}(\sum_{<\alpha,\beta> \in x} d_{\alpha\beta} + d_{ij}) + H^*(x \cup \{<i,j>\})] & \\ \hspace{6cm} \text{for } \mathcal{O}(x) \neq \emptyset & \\[2mm] 0 \quad \text{for } \mathcal{O}(x) = \emptyset & \end{cases} \quad (x \in \mathcal{X})$$

where, for simplicity, we have written $H^*(x)$ instead of $H^*(x, \sum_{<\alpha,\beta> \in x} d_{\alpha\beta})$.

Next, we sketch very briefly how to solve equation (5.10.4). Let K be the maximum number of arcs of any $x \in \mathcal{X}$:

$$K := \max_{x \in \mathcal{X}} |x|$$

Trivially, K is not greater than the number of arcs of the network. The sets

$$\mathcal{X}_k := \{x \in \mathcal{X} \mid |x| = k\}$$

can be generated successively for $k = 0, 1, \ldots, K$ beginning with $\mathcal{X}_0 = \{\emptyset\}$. To determine and address those sets in an efficient way, procedures for generating and labeling feasible sets of jobs subject to precedence constraints can be used, which have been developed by Schrage and Baker (1978) and by Lawler (1979) (compare also Rubach (1984)). After that, equation (5.10.4) can be solved for all $x \in \mathcal{X}_k$ successively for $k = K-1, \ldots, 0$ beginning with $H^*(x) = 0$ for $x \in \mathcal{X}_K$. For given x, the evaluation of equation (5.10.4) is simple.

If the GERT network in question contains cycles, the evaluation of Bellman's equation (5.10.3) becomes much more complicated because of the following facts:

(i)   The optimal scheduling policies are in general time–dependent even if the activity durations are deterministic (except that the objective function has the special form $\mathbb{E}(\Sigma wC)$, cf. (5.8.8)).

(ii)   The sets $x\in\mathfrak{X}$ generally contain arcs and nodes, and if we again introduce the sets

$$\mathfrak{X}_k := \{x\in\mathfrak{X}\,|\,x \text{ contains exactly } k \text{ arcs}\}$$

it may happen that both a set x and its follower $y_{ij}(x)$ owing to activity $\langle i,j\rangle$ belong to one and the same set $\mathfrak{X}_k$ (namely if activity $\langle i,j\rangle$ lies in a cycle). More precisely, it holds that

$$y\cap E \in \mathfrak{X}_k \cup \mathfrak{X}_{k+1} \quad \text{for all } y\in\mathcal{F}(x),\ x\in\mathfrak{X}_k$$

In the general case, Bellman's equation (5.10.3) can be evaluated by means of the so-called value–iteration and policy–iteration techniques of dynamic programming as we will see in section 6.4 (compare also Rubach (1984)).

Chapter 6 Cost Minimization for STEOR and EOR Networks

In this chapter we consider the case where different types of cost are incurred by the execution of activities and the occurrence of events of a project modelled by a STEOR network. We will see that the expected total cost of the project depends only on the activation functions Y_j of the individual nodes j of the network. Since the activation function Y_j of any node j in an admissible EOR network N coincides with the activation function of node j in each STEOR network from a covering of N that contains all nodes from $\mathcal{R}(j)$ and all arcs joining those nodes (cf. (3.5.10)), the cost minimization problem for admissible EOR networks of Markov degree d>1 can be solved in the same manner as for STEOR networks.

We assume in this chapter that the network in question has only one source (which is activated at time 0). Recall that if we have an admissible EOR network with several sources, the corresponding one–source network, which can be considered instead, also represents an admissible EOR network (cf. (1.3.4)).

The first model of optimal time–cost trade–offs in STEOR networks was offered by Arisawa and Elmaghraby (1972). In this model, the objective is to maximize the decrease in project duration per unit investment where the activity durations are the variables. This leads to a problem of fractional linear programming.

In what follows, we will discuss a more general cost minimization problem for STEOR networks with time–dependent execution probabilities and durations of activities and with several types of cost. To solve this problem we present a dynamic programming approach proposed by Nicolai (1982), where the decision variables are the time–dependent execution probabilities and distribution functions of the durations of the activities. Therefore, in section 6.1 we will first consider STEOR networks with time–dependent arc weights. A different approach to solving the cost minimization problem which leads to an optimal control problem was offered by Delivorias et al. (1984).

Cost minimization for general GERT networks seems to be very complicated. Delivorias (1990) deals with cost minimization in BES networks and proposes some solution methods. These algorithms, however, are very cumbersome and time–consuming and thus we have left them.

6.1 STEOR Networks with Time–Dependent Arc Weights

Let N be a STEOR network (with only one source) whose arc weights $\begin{bmatrix} p_{ij} \\ F_{ij} \end{bmatrix}$ depend on the time of activation of node i, that is,

$$p_{ij}(s) := P(<i,j> \text{ is carried out} \mid i \text{ has occurred at time } s)$$

$$F_{ij}(s,t) := \begin{cases} P(D_{ij}^{\alpha} \leq t \mid \alpha\text{th execution of } <i,j> \text{ has been} \\ \quad \text{begun at time } s) \qquad \text{for } t \geq 0 \\ 0 \quad \text{for } t < 0 \end{cases} \qquad (s \in \mathbb{R}_+)$$

where D_{ij}^{α} is the duration of the αth execution of activity $<i,j>$. Again, $p_{ij}(\cdot)$ and $F_{ij}(\cdot,\cdot)$ are supposed to be independent of how many times project event i has occurred or, respectively, activity $<i,j>$ has been carried out before, and we speak of the duration D_{ij} of activity $<i,j>$ (cf. (1.2.5)). For all $<i,j> \in E$ and $t \in \mathbb{R}_+$, the functions $p_{ij}(\cdot)$ and $F_{ij}(\cdot,t)$ are assumed to be measurable on $\mathbb{R}_+$ [21].

As in chapters 1 to 4 we stipulate that each activity is begun at its earliest possible start time. In assumption A2c, the random variables D_{ij}^{α} and B_i^{β} now depend on the time of activation of node i. Assumption A3 has to be replaced by

Assumption $\overline{A3}$.

For each node k of a cycle structure C, there are an $\epsilon > 0$ and a path from k to a node outside C such that $p_{ij}(s) \geq \epsilon$ for every arc $<i,j>$ of this path and for all $s \geq 0$.

---

[21]   In what follows, we will speak of a set or a function as measurable without explicitly mentioning the σ–algebras involved. Of course, measurable on $\mathbb{R}$ or on $\mathbb{R}_+$ means Borel–measurable. The reader who is unfamiliar with the elements of measure theory and integration may contact Rohatgi (1976), introduction, and Apostol (1974), chapter 10, or may think of Riemann–integrable functions instead of measurable ones.

(6.1.1) We again introduce the expanded STEOR network N^+ with node set V^+ in the same way as in (3.2.1). Moreover, let X_ν denote that node of N^+ whose activation is the νth node activation in turn and let θ_ν be the time of that activation ($\nu \in \mathbb{N}_0$), where X_0 is the source and $\theta_0 = 0$. Then $(X_\nu, \theta_\nu)_{\nu \in \mathbb{N}_0}$ represents a nonstationary Markov renewal process with state space $V^+ \times \mathbb{R}_+$ and transition functions Q_{ij} given by

$$Q_{ij}(s,t) := \begin{cases} P(X_{\nu+1}=j, \theta_{\nu+1} \leq s+t \mid X_\nu=i, \theta_\nu=s) & \text{for } t \geq 0 \\ 0 & \text{for } t < 0 \end{cases} \qquad (i,j \in V^+; s \geq 0)$$

(compare (3.1.7)) which are independent of ν. It holds that

$$Q_{ij}(s,t) = p_{ij}(s) F_{ij}(s,t)$$

(cf. theorem (3.2.4)). The Markov renewal process $(X_\nu, \theta_\nu)_{\nu \in \mathbb{N}_0}$ is said to be **nonstationary** because the transition functions Q_{ij} depend on the time of activating node i. The μ–step transition functions $Q_{ij}^{(\mu)}$ are defined by

$$Q_{ij}^{(\mu)}(s,t) := \begin{cases} P(X_{\nu+\mu}=j, \theta_{\nu+\mu} \leq s+t \mid X_\nu=i, \theta_\nu=s) & \text{for } t \geq 0 \\ 0 & \text{for } t < 0 \end{cases}$$

$$(i,j \in V^+; \mu \in \mathbb{N}, s \geq 0)$$

(cf. (3.1.9)) and are independent of ν. Similarly to the case of stationary Markov renewal processes, the μ–step transition functions can be computed successively beginning with $Q_{ij}^{(1)} = Q_{ij}$ (compare (3.1.11) and Nicolai (1980)).

The activation functions Y_j of the STEOR network N, which are defined in the same way as for networks with time–independent arc weights, satisfy the system of integral equations

$$(6.1.2) \quad \left\{ \begin{array}{l} Y_1(t)=1 \\ Y_j(t)=p_{1j}(0)F_{1j}(0,t)+ \sum_{k=2}^{n} \int_{[0,t]} p_{kj}(s)F_{kj}(s,t-s)Y_k(ds) \\ \hspace{6cm} (j=2,\ldots,n) \end{array} \right\} \quad (t \geq 0)$$

where $V=\{1,\ldots,n\}$ is the node set of N and node 1 is the source (compare (3.5.3)). The system of integral equations (6.1.2) can be solved approximately in the same manner as sketched in section 3.5. For additional details of STEOR networks with time–dependent arc weights and the corresponding Markov renewal processes we refer to Nicolai (1980).

Note that the activation functions Y_{ij} $(i\in R, j\in \mathcal{R}(i))$ for STEOR networks with time–dependent arc weights and source set R $(|R|>1)$ additionally depend on the time of activation of source i, that is, definition (3.5.11) for $t\geq 0$ has to be replaced by

$$Y_{ij}(s,t):=\mathbb{E}[K_j(s+t)|T_i=s]$$

The activation functions Y_{ij}, however, are not needed in what follows because we have assumed that the underlying network has only one source.

6.2 Cost Minimization in STEOR Networks: Basic Concepts

We suppose that the arc weights p_{ij} and F_{ij} are not fixed in advance but may be affected by certain actions. Furthermore, there are in general costs incurred by the execution of the project, which depend on the actions selected. For example, we may hire more labour or use additional machinery to reduce the duration D_{ij} of an activity $<i,j>$ (that is, we "shift" the distribution function F_{ij} to the left) or to increase the execution probability of an activity that leads to a sink representing the successful completion of the project. In either case, an additional cost associated with the respective activity arises by taking such an action and, to have less cost, we should choose a greater activity duration or smaller execution probability of the activity, respectively. On the other hand, a penalty cost depending on the duration of the total project or the probability of a successful project termination might be incurred so that we are interested in smaller activity durations or larger execution probabilities of certain activities, respectively. The objective is to choose the actions or, respectively, arc weights p_{ij} and F_{ij} so as to minimize the expected total cost of the project.

(6.2.1) Now suppose that when a project event i occurs at time $s\geq 0$, an **action** $\delta(i,s)$ from a nonempty measurable action set $\Delta(i,s)\subseteq R$ can be executed. The weights $p_{ij}(s)$ and $F_{ij}(s,\cdot)$ of the arcs $<i,j>$ emanating from node i are assumed to be specified only after the selection of such an action (in other words, the arc weights depend on that action). A measurable function $\delta:V\times\mathbb{R}_+\rightarrow\mathbb{R}$, which assigns an action $\delta(i,s)\in\Delta(i,s)$ to each state $(i,s)\in V\times\mathbb{R}_+$, is called a **policy**. Let Δ denote the set of all such policies.

The selection of a policy $\delta\in\Delta$ implies the specification of the weights of all arcs of the STEOR network (with time–dependent arc weights) in question. The network is then denoted by N_δ to order to indicate that its arc weights depend on policy δ. Henceforth, the *assumptions A2* and $\overline{A3}$ *are supposed to be satisfied for all policies* $\delta\in\Delta$. In particular, the positive number ϵ in $\overline{A3}$ is to be independent of the policy δ chosen.

(6.2.2) The quantities related to the STEOR network N_δ are denoted by the superscript δ (or $\delta(i,s)$, respectively). Examples are the weights $p_{ij}^\delta(s)$ and $F_{ij}^\delta(s,t)$ (or $p_{ij}^{\delta(i,s)}(s)$ and $F_{ij}^{\delta(i,s)}(s,t)$, respectively), the activation functions Y_j^δ and activation numbers z_j^δ as well as the Markow renewal process $(X_\nu,\theta_\nu)_{\nu\in\mathbb{N}_0}^\delta$ and its transition functions Q_{ij}^δ and μ–step transition functions $(Q_{ij}^\delta)^{(\mu)}$. The set $\{(X_\nu,\theta_\nu)_{\nu\in\mathbb{N}_0}^\delta\,|\,\delta\in\Delta\}$ is also called a **Markow renewal decision process**, where the word "decision" indicates that we will be looking for a policy δ that minimizes a certain objective function.

(6.2.3) Next, we introduce several **types of cost**:

(i)   The **event cost** $c_e(i,s)$ results from the occurrence of event i at time s (for instance, a cost depending on the project duration or a penalty cost in case a prescribed occurrence time of event i is exceeded).

(ii)   The **activity cost** $c_{ij}(s,t)$ is incurred when the execution of activity $<i,j>$ is begun at time s and is terminated at time s+t (for example, a cost depending on the (realized) duration t of activity $<i,j>$).

(iii)   The **action cost** $c_a^{\delta(i,s)}(i,s)$ arises when event i occurs at time s and action $\delta(i,s)$ is taken.

The cost functions $c_{ij}(\cdot,\cdot)$ $(<i,j>\in E)$, $c_e(i,\cdot)$ and $c_a^{\delta(i,\cdot)}(i,\cdot)$ $(i\in V)$ are assumed to be measurable and bounded on $\mathbb{R}_+^2$ or $\mathbb{R}_+$, respectively. For each sink i and all $s\in\mathbb{R}_+$, the action cost $c_a^{\delta(i,s)}(i,s)$ is to be equal to 0. We suppose that there is no terminal cost when the project has been completed.

Let C_δ be the expected total cost of the project if policy δ is selected. C_δ results from summing up the expected costs of types (i), (ii) and (iii) over all nodes and arcs, respectively, of the STEOR network as follows:

Expected event cost of event i:

(6.2.4) $\qquad C_e^\delta(i) = \int_{\mathbb{R}_+} c_e(i,s)\, Y_i^\delta(ds)$

Expected activity cost of activity $<i,j>$:

$$C^\delta(i,j) = \int_{\mathbb{R}_+} [\int_{\mathbb{R}_+} c_{ij}(s,t)\, F_{ij}^\delta(s,dt)]\, p_{ij}^\delta(s)\, Y_i^\delta(ds)$$

Expected action cost for event i:

$$C_a^\delta(i) = \int_{\mathbb{R}_+} c_a^{\delta(i,s)}(i,s)\, Y_i^\delta(ds)$$

Then

(6.2.5) $\qquad C_\delta = \sum_{i\in V} [C_e^\delta(i) + C_a^\delta(i)] + \sum_{<i,j>\in E} C^\delta(i,j)$

and the optimization problem to be solved becomes

(6.2.6) $\qquad \left\{ \begin{array}{l} \text{Minimize} \quad C_\delta \\ \text{subject to } \delta\in\Delta \end{array} \right.$

Our cost model includes a large number of special problems which need not originally represent cost minimization problems. For example, if we wish to minimize the expected total duration of the project under consideration, we choose the following special event cost:

(6.2.7) $c_e(i,s) := \begin{cases} s & \text{if } i \in S \\ 0, & \text{otherwise} \end{cases}$ $(s \geq 0)$

where S is again the sink set. All remaining cost functions are equal to 0. Then (6.2.5), (6.2.4), (6.2.7) and (2.3.3) provide

$$C_\delta = \sum_{i \in V} c_e^\delta(i) = \sum_{i \in S} \int_{\mathbb{R}_+} s Y_i^\delta(ds) = \int_{\mathbb{R}_+} s G_\delta(ds)$$

where G_δ is the distribution function of the duration of the project corresponding to the STEOR network N_δ in question (which coincides with the distribution function of the skipping project duration). Thus, C_δ equals the expected project duration when policy δ is chosen.

Sometimes some of the terminal events of the project that represent a successful project completion are of particular interest, say, all sinks $i \in S' \subset S$. If we want to maximize the probability that one of those terminal events occurs, we choose the following event cost:

$c_e(i,s) := \begin{cases} -1 & \text{if } i \in S' \\ 0, & \text{otherwise} \end{cases}$ $(s \geq 0)$

Again all remaining cost functions are equal to 0. Then we have

$$C_\delta = - \sum_{i \in S'} \int_{\mathbb{R}_+} Y_i^\delta(ds) = - \sum_{i \in S'} z_i^\delta$$

where $\sum_{i \in S'} z_i^\delta$ is equal to the probability that one of the sinks $i \in S'$ of the STEOR network N_δ in question is activated.

Note that these two special cases assume that the one–sink condition is satisfied and thus merely apply to STEOR networks (with only one source).

(6.2.8) The scheduling problem $1|\text{GERT},D\!\sim\!G|\mathbb{E}(\Sigma f)$ for an admissible GERT network N (with only one source) from chapter 5 can also be viewed as a cost minimization problem for an appropriate STEOR network. Given a scheduling policy ϕ, we consider the STEOR network N_ϕ with node set $\mathfrak{X}$ (the set of the feasible arc–node sets) and arc set $\{<x,y>\,|\,x\in\mathfrak{X},y\in\mathcal{F}(x)\}$. To specify the arc weights in N_ϕ we note that to each arc $<x,y>$ in N_ϕ where $y=y_{ij}(x)$ is the follower of x owing to activity $<i,j>$, there corresponds exactly one arc in N, namely the arc $<i,j>$. Then the arc weights in N_ϕ are given by

$$p^\phi_{xy}(s):=\begin{cases} p_{ij} & \text{if } y=y_{ij}(x) \text{ and } <i,j>\in\phi(x,s) \\ 1 & \text{if } O(x)=\emptyset \text{ and } x=y \qquad\qquad (x,y\in\mathfrak{X};s\geq0) \\ 0, & \text{otherwise} \end{cases}$$

$$F^\phi_{xy}(s,t):=\begin{cases} F_{ij}(t) & \text{if } y=y_{ij}(x) \text{ and } <i,j>\in\phi(x,s) \\ 1 & \text{if } O(x)=\emptyset \text{ and } x=y \qquad\qquad (x,y\in\mathfrak{X};s,t\geq0) \\ 0, & \text{otherwise} \end{cases}$$

where $\begin{bmatrix}p_{ij}\\F_{ij}\end{bmatrix}$ are the arc weights in N (compare (5.10.1) and (5.8.11)). Since a cost $f_{ij}(s+t)$ is incurred when an execution of activity $<i,j>$ in N is begun at time s and terminated at time $s+t$, we have an activity cost

(6.2.9) $$c_{xy}(s,t):=\begin{cases} f_{ij}(s+t) & \text{if } y=y_{ij}(x) \\ 0, & \text{otherwise} \end{cases} \qquad (x,y\in\mathfrak{X};s,t\geq0)$$

The remaining cost functions are equal to 0.

6.3 A Dynamic Programming Approach

In this section we present a dynamic programming approach to the cost minimization problem (6.2.6) in a STEOR network N (which can also be applied to admissible EOR networks as mentioned at the beginning of chapter 6). Note that a special case of the following model (with activity costs of the form (6.2.9) and vanishing event and action costs) has already been discussed in section 5.10.

If node i is activated at time s and action $\delta(i,s)$ is taken, the **expected single-stage cost** incurred by the transition from node i to any successor becomes

(6.3.1)
$$h^{\delta(i,s)}(i,s) = c_e(i,s) + c_a^{\delta(i,s)}(i,s)$$

$$+ \sum_{j\in S(i)} p_{ij}^{\delta(i,s)}(s) \int_{\mathbb{R}_+} c_{ij}(s,t) F_{ij}^{\delta(i,s)}(s,dt)$$

(compare (5.10.2)). Let $H_\delta(i,s)$ be the expected total cost of the "partial project" starting at state (i,s) if policy δ is chosen. The corresponding mapping $H_\delta : V\times\mathbb{R}_+\to\mathbb{R}$ is called the **total cost function** for policy $\delta\in\Delta$. In Nicolai (1982) it is shown that H_δ can be expressed in terms of h^δ and the μ–step transition functions of the Markov renewal process $(X_\nu,\theta_\nu)_{\nu\in\mathbb{N}_0}^\delta$ as follows:

(6.3.2) $$H_\delta(i,s) = \sum_{\mu=0}^{\infty} \sum_{j\in V} \int_{\mathbb{R}_+} h^{\delta(j,s+t)}(j,s+t)(Q_{ij}^\delta)^{(\mu)}(s,dt) \quad ((i,s)\in V\times\mathbb{R}_+)$$

Moreover, it is shown that the family of functions $\{H_\delta | \delta\in\Delta\}$ is uniformly bounded.

We wish to minimize the expected total cost of the project. Hence, we seek to determine the **minimum total cost function** H^* given by

$$H^*(i,s) := \inf_{\delta\in\Delta} H_\delta(i,s) \quad \text{for } (i,s)\in V\times\mathbb{R}_+$$

and an **optimal policy** $\delta^*\in\Delta$ with $H_{\delta^*}=H^*$. Let node 1 be again the source of the network. Since the project starts at state $(1,0)$, $H^*(1,0)$ is the infimum expected total cost of the project.

The problem of finding the minimum total cost function H^* and a corresponding optimal policy δ^* represents a stochastic dynamic programming problem (compare sections 5.9 and 5.10). **Bellman's equation** now has the form

$$(6.3.3) \qquad H^*(i,s) = \inf_{\delta(i,s)\in\Delta(i,s)} \{h^{\delta(i,s)}(i,s) +$$

$$+ \sum_{j\in S(i)} p_{ij}^{\delta(i,s)}(s) \int_{\mathbb{R}_+} H^*(j,s+t) F_{ij}^{\delta(i,s)}(s,dt)\}$$

$$\text{for } (i,s)\in V\times\mathbb{R}_+$$

(cf. (5.10.3)).

Bellman's equation (6.3.3) can be solved by means of the so-called **value-iteration** and **policy-iteration methods**. In what follows, we will describe those two techniques in a fairly abstract form using some concepts from functional analysis (most of the subsequent material can be found in Denardo (1967) and Heyman and Sobel (1984), section 5.4; for the basic concepts from functional analysis we refer to Apostol (1974), chapters 3 and 4, and Dieudonné (1969), chapter 3). That approach also includes the verification of some results stated thus far without proof, for instance, the fact that the minimum total cost function H^* is the unique solution to Bellman's equation (6.3.3).

H^* is a bounded function that maps the state space $V\times\mathbb{R}_+$ [22] into the set of real numbers $\mathbb{R}$. We denote the set of all bounded functions $f:V\times\mathbb{R}_+\to\mathbb{R}$ by B. Next, we introduce a **distance** $\rho(f_1,f_2)$ for every two such functions f_1 and f_2 by

$$(6.3.4) \qquad \rho(f_1,f_2) := \sup_{(i,s)\in V\times\mathbb{R}_+} |f_1(i,s)-f_2(i,s)|$$

ρ has the following three defining properties of a distance:

---

[22]   Although $V^+\times\mathbb{R}_+$ is the state space of the underlying Markov renewal decision process, sometimes we also refer to $V\times\mathbb{R}_+$ as the state space.

(6.3.4a) $\rho(f_1,f_2)=0$ exactly if $f_1=f_2$

(6.3.4b) $\rho(f_1,f_2)=\rho(f_2,f_1)$

(6.3.4c) $\rho(f_1,f_2)\leq\rho(f_1,g)+\rho(g,f_2)$ for all $g\in B$ (triangle inequality)

The set of functions B together with the distance ρ forms a so-called **metric space**. This metric space is *complete*, that is, each Cauchy sequence [23] in B converges to an element of B (cf. Apostol (1974), sections 4.3 and 4.4).

A mapping $L:B\to B$ of the metric space B into itself is also called an **operator**. For $\delta\in\Delta$, we introduce the operator L_δ by

(6.3.5) $$L_\delta(f)(i,s) := h^{\delta(i,s)}(i,s) +$$

$$+ \sum_{j\in S(i)} p_{ij}^{\delta(i,s)}(s) \int_{\mathbb{R}_+} f(j,s+t)F_{ij}^{\delta(i,s)}(s,dt)$$

$$\text{for } f\in B, (i,s)\in V\times\mathbb{R}_+$$

Note that $L_\delta(f)$ is an element of B, that is, a function defined on the state space $V\times\mathbb{R}_+$. $L_\delta(f)(i,s)$ then means the value of this function at (i,s). Since the family of functions $\{L_\delta(f)\,|\,\delta\in\Delta\}$ is uniformly bounded for each fixed $f\in B$, we may introduce the **minimum operator** $L^*:B\to B$ by

(6.3.6) $$L^*(f)(i,s) := \inf_{\delta\in\Delta} L_\delta(f)(i,s) \quad \text{for } f\in B, (i,s)\in V\times\mathbb{R}_+$$

f is called a **fixed point** of an operator L if L maps f into itself, that is, $f=L(f)$. Since Bellman's equation (6.3.3) can be rewritten as $H^*=L^*(H^*)$, the minimum total cost function H^* is a fixed point of the minimum operator L^*.

---

[23] (f_ν) is said to be a Cauchy sequence if, for each $\eta>0$, there is a $k(\eta)$ such that $\rho(f_\mu,f_\nu)<\eta$ for all $\mu,\nu\geq k(\eta)$.

Next, we consider some properties of the operators L_δ and L^*. An operator L is called isotone (or nondecreasing) if

$$L(f_1) \leq L(f_2) \quad \text{whenever } f_1 \leq f_2$$

where $f_1 \leq f_2$ means that $f_1(i,s) \leq f_2(i,s)$ for all $(i,s) \in V \times \mathbb{R}_+$. The next proposition follows immediately from the definitions (6.3.5) and (6.3.6) of the operators L_δ and L^*.

(6.3.7)   Proposition.
The operators L_δ ($\delta \in \Delta$) and L^* are isotone.

The **modulus** of an operator $L:B \to B$ is the smallest number α such that

$$\rho(L(f_1), L(f_2)) \leq \alpha \rho(f_1, f_2) \quad \text{for } f_1, f_2 \in B$$

If modulus $\alpha < 1$, the operator L is said to be a **contraction**.

Let r be the so–called **path rank** of the dummy sink of the expanded STEOR network N^+ in question, that is, the maximum number of arcs of all paths in N^+ whose final node is the dummy sink. Trivially, $r \leq n$ where n is again the number of nodes in N.

Let L^k be the kth power of operator L, which is defined recursively by $L^k(f) := L(L^{k-1}(f))$ for $k > 1$. Using assumption $\overline{A3}$ the following proposition can be proved (cf. Nicolai (1982)):

(6.3.8)   Proposition.
For each $\delta \in \Delta$, the operator L_δ has modulus 1 and L_δ^r has modulus $1 - \epsilon^r$, where $\epsilon > 0$ is given by assumption $\overline{A3}$.

The so–called **fixed–point theorem for k–stage contractions** (see Denardo (1967)) says that if L is a mapping of a complete metric space into itself and L^k has modulus < 1 for some positive integer k, then L has a unique fixed point. Since B is complete and by proposition (6.3.8) L_δ^r has modulus <1, L_δ has a unique fixed point. Moreover, it can be shown (cf. Nicolai (1982))

(6.3.9) Proposition.
For each $\delta\in\Delta$, the total cost function H_δ is the unique fixed point of operator L_δ.

Using propositions (6.3.7), (6.3.8) and (6.3.9) the following theorem can be proved (cf. Denardo (1967)).

(6.3.10)   Theorem.
   (a) If $f\geq L_\delta(f)$, then $L_\delta(f)\geq H_\delta$ $(f\in B, \delta\in\Delta)$
   (b) H^* is the unique fixed point of operator L^*
   (c) If $f\geq H^*$, then $\rho(L^{*r}(f),H^*)\leq(1-\epsilon^r)\rho(f,H^*)$ $(f\in B)$

Note, that (6.3.10b) says that the minimum total cost function H^* is the unique solution to Bellman's equation (6.3.3).

By definition of L^* and proposition (6.3.9) we have $L^*(H_\delta)\leq L_\delta(H_\delta)=H_\delta$. (6.3.10b), the definition of H^*, and the monotonicity of L^* provide $H^*=L^*(H^*)\leq L^*(H_\delta)$. Hence,

$$H^*\leq L^*(H_\delta)\leq H_\delta \quad \text{for all } \delta\in\Delta$$

and thus by (6.3.4)

(6.3.11) $\rho(L^*(H_\delta),H^*)\leq\rho(H_\delta,H^*)$ for all $\delta\in\Delta$

To guarantee the existence of an optimal policy δ^*, we require in what follows that there is a nonempty subset B' of B such that

(6.3.12) $H_\delta, H^*, L_\delta(f)$, and $L^*(f)\in B'$ for all $\delta\in\Delta, f\in B'$

(6.3.13)   For each $f\in B'$, there exists a policy $\gamma\in\Delta$ such that

$$L_\gamma(f) = \inf_{\delta \in \Delta} L_\delta(f) = L^*(f)$$

The conditions (6.3.12) and (6.3.13) are satisfied, for example, if all action sets $\Delta(i,s)$, $(i,s) \in V \times \mathbb{R}_+$, are finite (we then have $B' = B$) or if these sets are compact, and the cost functions c_{ij}, $c_e(i,\cdot)$, and $c_a^{\delta(i,\cdot)}(i,\cdot)$ as well as the functions $p_{ij}(\cdot)$ and $F_{ij}(\cdot,t)$ satisfy certain continuity conditions. The existence of an optimal policy permits us to write "min" instead of "inf" henceforth.

In the next section, we will discuss the value–iteration and policy–iteration methods, which provide approximations for the minimum total cost function H^* and an optimal policy δ^*.

6.4 The Value–Iteration and Policy–Iteration Techniques

From (6.3.10c) with $f=H_\delta$ and (6.3.11) it follows that

$$\lim_{\nu \to \infty} \rho(L^{*\nu}(H_\delta), H^*) = 0 \quad \text{for all } \delta \in \Delta$$

Hence, if we start the so–called method of successive approximations, that is, the recursion

$$H^{(\nu+1)} := L^*(H^{(\nu)}) \quad \text{for } \nu=0,1,2,\ldots$$

with $H^{(0)} := H_\delta$, where δ is any policy, then

$$\lim_{\nu \to \infty} \rho(H^{(\nu)}, H^*) = 0$$

In other words, the sequence $(H^{(\nu)})$ obtained by repeatedly applying the minimum operator L^* converges to the minimum total cost function H^*. This gives the following

(6.4.1)   Algorithm (Successive approximations).

Step 1.   Choose an arbitrary policy $\delta \in \Delta$ and set $H:=H_\delta$

Step 2.   Compute $L^*(H):=\min\limits_{\delta \in \Delta} L_\delta(H)$ and $\gamma \in \Delta$ such that $L_\gamma(H)=\min\limits_{\delta \in \Delta} L_\delta(H)$

Step 3.   (Stopping rule). If $m_H \leq \eta$, compute the solution H_γ to the fixed–point equation
$f=L_\gamma(f)$ and terminate;

otherwise set $H:=L^*(H)$ and go to step 2   ∎

$\eta>0$ represents a prescribed tolerance and m_H is a stopping quantity depending on function H, for example,

$$(6.4.2) \qquad m_H := \sup_{(i,s)\in V\times\mathbb{R}_+} \left| \frac{H(i,s) - L^*(H)(i,s)}{L^*(H)(i,s)} \right|$$

After termination of the algorithm, $H_\gamma(1,0)$ is an approximation of the minimum expected total project cost and γ is an approximation of an optimal policy.

Let $H^{(0)}$ be again the initial approximation of H^* and $H^{(\nu)}:=L^{*\nu}(H^{(0)})$ for $\nu \in \mathbb{N}$. Then a bound on $\rho(H^{(\nu)},H^*)$ can be found as follows (compare Denardo (1967)): Using (6.3.4) and proposition (6.3.8) we obtain for $\nu \in \mathbb{N}$

$$(6.4.3) \qquad \rho(H^{(\nu)},H^*) \leq \sum_{\mu=1}^{\infty} \rho(H^{(\nu+\mu)},H^{(\nu+\mu-1)})$$

$$\leq r\rho(H^{(\nu)},H^{(\nu-1)}) \sum_{\mu=0}^{\infty} (1-\epsilon^r)^\mu = \frac{r}{\epsilon^r} \rho(H^{(\nu)},H^{(\nu-1)})$$

Trivially,

$$\rho(H^{(\nu)},H^*) \leq \rho(H^{(\mu)},H^*) \text{ for } \nu \in \mathbb{N} \text{ and } \mu=0,1,\ldots,\nu-1$$

and by (6.3.10c)

$$\rho(H^{(\mu+r)},H^*) \le (1-\epsilon^r)\rho(H^{(\mu)},H^*) \quad \text{for } \mu \in \mathbb{N}_0$$

Thus, for $\nu \in \mathbb{N}$

$$\rho(H^{(\nu)},H^*) \le (1-\epsilon^r)^{\left\lfloor \frac{\nu-\mu}{r} \right\rfloor}\rho(H^{(\mu)},H^*) \quad \text{for } \mu=0,1,\ldots,\nu-1$$

or

(6.4.4) $$\rho(H^{(\nu)},H^*) \le \min_{\mu=0,1,\ldots,\nu-1} (1-\epsilon^r)^{\left\lfloor \frac{\nu-\mu}{r} \right\rfloor}\rho(H^{(\mu)},H^*)$$

where $\lfloor a \rfloor$ is the greatest integer $\le$ a. From (6.4.3) and (6.4.4) it follows: Let

$$a_1 := \frac{r}{\epsilon^r}\rho(H^{(1)},H^{(0)})$$

$$a_\nu := \min \left[\frac{r}{\epsilon^r}\rho(H^{(\nu)},H^{(\nu-1)}),(1-\epsilon^r)^{\beta_\nu} a_{\nu-1}\right] \quad \text{for } \nu=2,3,\ldots$$

$$\text{where } \beta_\nu := \begin{cases} 1 \text{ if } \nu \text{ is an integral multiple of r} \\ 0, \text{ otherwise} \end{cases}$$

Then

$$\rho(H^{(\nu)},H^*) \le a_\nu \text{ for } \nu \in \mathbb{N}$$

If ϵ is small, the bounds a_ν are very poor.

Since the method of successive approximations determines a sequence of quantities $L^{*\nu}(H)$ referred to as *values* in dynamic programming, it is also called the **value–iteration** technique. The following **policy–iteration** method or **policy–improvement technique** is aimed at the construction of a sequence of *policies*.

Given $\gamma \in \Delta$, let $\gamma^+ \in \Delta$ be such that

(6.4.5) $$L_{\gamma^+}(H_\gamma) = \min_{\delta \in \Delta} L_\delta(H_\gamma) = L^*(H_\gamma)$$

Since H_γ is the fixed point of L_γ, (6.4.5) yields

$$H_\gamma = L_\gamma(H_\gamma) \geq L_{\gamma^+}(H_\gamma)$$

This relation and (6.3.10a) with $f=H_\gamma$ and $\delta=\gamma^+$ provide

(6.4.6) $$H_\gamma \geq L_{\gamma^+}(H_\gamma) \geq H_{\gamma^+}$$

In other words, policy γ^+ is better than or at least as good as policy γ. In this manner, a sequence $\gamma_0, \gamma_1, \ldots$ of policies can be determined such that the sequence of the corresponding total cost functions $H_{\gamma_0}, H_{\gamma_1}, \ldots$ is nonincreasing. From

$$H^* \leq H_{\gamma_{\nu+1}} \leq H_{\gamma_\nu} \quad \text{for } \nu \in \mathbb{N}_0$$

and (6.3.4) it follows that

(6.4.7) $$\rho(H_{\gamma_{\nu+1}}, H^*) \leq \rho(H_{\gamma_\nu}, H^*) \quad \text{for } \nu \in \mathbb{N}_0$$

By (6.4.6) and (6.4.5) with $\gamma=\gamma_\nu$ and $\gamma^+=\gamma_{\nu+1}$ we have

$$H_{\gamma_{\nu+1}} \leq L^*(H_{\gamma_\nu})$$

Applying this relation r times and observing the monotonicity property of operator L^* (proposition (6.3.7)) we obtain

$$H^* \leq H_{\gamma_{\nu+r}} \leq L^{*r}(H_{\gamma_\nu})$$

and by (6.3.4) and (6.3.10c)

$$(6.4.8) \qquad \rho(H_{\gamma_{\nu+r}}, H^*) \le \rho(L^{*r}(H_{\gamma_{\nu}}), H^*) \le (1-\epsilon^r)\rho(H_{\gamma_{\nu}}, H^*) \quad \text{for } \nu \in \mathbb{N}_0$$

(6.4.7) and (6.4.8) yield the convergence of the sequence $(H_{\gamma_{\nu}})$ to H^*:

$$(6.4.9) \qquad \lim_{\nu \to \infty} \rho(H_{\gamma_{\nu}}, H^*) = 0$$

This gives

(6.4.10)   Algorithm (Policy iteration).
Step 1.   Choose an arbitrary policy $\gamma \in \Delta$
Step 2.   Compute the solution H_{γ} to the fixed–point equation $f = L_{\gamma}(f)$
Step 3.   (Stopping rule). If $m_{H_{\gamma}} \le \eta$, terminate;

otherwise find $\gamma^+ \in \Delta$ such that $L_{\gamma^+}(H_{\gamma}) = \min_{\delta \in \Delta} L_{\delta}(H_{\gamma})$, set $\gamma := \gamma^+$, and go to

step 2

■

Again $\eta > 0$ is a prescribed tolerance and m_H is a stopping quantity, for instance, given by (6.4.2).

The policy–improvement technique constructs a sequence of policies (γ_{ν}) such that the corresponding sequence $(H_{\gamma_{\nu}})$ is nonincreasing. Since

$$H_{\gamma_{\nu+1}} \le L^*(H_{\gamma_{\nu}}) = \min_{\delta \in \Delta} L_{\delta}(H_{\gamma_{\nu}})$$

the determination of an improved policy γ^+ in step 3 of algorithm (6.4.10) can be replaced by the following step:

Compute $H := L^{*k}(H_{\gamma})$ for some $k \in \mathbb{N}$ and $\gamma^+ \in \Delta$ such that $L_{\gamma^+}(H) = L^*(H)$

Setting k>1 might improve the speed of the algorithm in case that the computation of $\min_{\delta\in\Delta} L_\delta(H_\gamma)$ takes less time than finding H_γ, that is, solving a fixed–point equation.

We see immediately that the computation time needed for one iteration step of the value–iteration method is less than for one iteration step of the policy–improvement technique because the latter method requires the solution of a fixed–point equation in each iteration in addition to the minimization. On the other hand, the number of iteration steps required by the policy–iteration method may be smaller than for the successive–approximation method. Moreover, the policy–improvement technique terminates as soon as two successive policies γ_ν and $\gamma_{\nu+1}$ coincide. If $\gamma_{\nu+1}=\gamma_\nu$ and thus $L_{\gamma_{\nu+1}}=L_{\gamma_\nu}$ for some $\nu\in\mathbb{N}_0$, then by proposition (6.3.9) $H_{\gamma_{\nu+1}}=H_{\gamma_\nu}$ and the policy iteration provides $H_{\gamma_\mu}=H_{\gamma_\nu}$ for all $\mu\geq\nu$ and by (6.4.9) $\rho(H_{\gamma_\nu},H^*)=0$. (6.3.4a) then gives $H_{\gamma_\nu}=H^*$. This case always occurs if both the set of all policies Δ and the state space are finite [24]. The value–iteration method, on the other hand, terminates if $H^{(\nu+1)}=H^{(\nu)}$ for some $\nu\in\mathbb{N}_0$ but generally not if $\gamma^{(\nu+1)}=\gamma^{(\nu)}$ (where, for $\mu=\nu,\nu+1$ $\gamma^{(\mu)}$ is such that $L_{\gamma^{(\mu)}}(H^{(\mu)})=L^*(H^{(\mu)})$).

Next, we deal with some computational aspects of solving our cost minimization problem. To compute $L_\delta(H)$ for given δ and H numerically (compare (6.3.5)), the time domain $\mathbb{R}_+$ has to be replaced by a bounded interval. In practice, a project often has to be completed by a prescribed point in time, otherwise it will be discontinued. In theory, for admissible EOR networks with time–dependent arc weights it holds that $P(D^{skip}=D<\infty)=1$, where D and D^{skip} are again the project duration and skipping project duration, respectively. We now state a condition which ensures that, for each admissible EOR network, there is a time $\bar{t}\in\mathbb{R}_+$ such that $P(D\leq\bar{t})=1$.

---

[24]   Later we will see that the time domain $\mathbb{R}_+$ can be replaced by a compact interval

$[0,\bar{t}]$. If, in addition, we only consider discrete time points in numerical computation, then the state space $V\times\mathbb{R}_+$ reduces to a finite set and the fixed–point equation $f=L_\gamma(f)$ becomes a system of finitely many linear equations.

(6.4.11) Condition.

(a) For each arc $\langle i,j \rangle$ of the network, there exists a $\hat{t}_{ij} \geq 0$ such that $F_{ij}^{\delta(i,s)}(s,\hat{t}_{ij})=1$ for every $\delta(i,s)\in\Delta(i,s)$ and all $s\geq0$.

(b) There exists an $\hat{s}\geq0$ such that for each node k of any cycle structure C, there is a path from k to a node outside C with $p_{ij}^{\delta(i,j)}(s)=1$ for every arc $\langle i,j \rangle$ of this path, every $\delta(i,s)\in\Delta(i,s)$, and all $s\geq\hat{s}$.

(6.4.11a) says that each activity has bounded duration. Condition (6.4.11) guarantees that, with probability 1, there is a $\tilde{t}\geq0$ such that each cycle structure is left, at the latest, a time period $\tilde{t}$ after the beginning of the project independently of the policy selected [25]. More specifically,

$$P(D\leq\bar{t})=1 \quad \text{for } \bar{t}:=\hat{s}+ \sum_{\langle i,j\rangle\in E} \hat{t}_{ij}$$

Thus, in (6.3.1), (6.3.2), (6.3.3), and (6.3.5), the integration over $\mathbb{R}_+$ can be replaced by the integral over the bounded interval $[0,\bar{t}-s]$.

Now let $V=\{1,\ldots,n\}$ be again the node set of the network in question. The (approximate) determination of the minimum total cost function H^* and an optimal policy δ^* requires much less computational effort if the following assumption $\overline{A0}$, which is the analogue to assumption A0 from section 3.6, is satisfied.

Assumption $\overline{A0}$.

There is a $\theta>0$ such that $F_{ij}^{\delta(i,s)}(s,\theta)=0$ for every arc $\langle i,j \rangle$ with $i\geq j$, for every $\delta(i,s)\in\Delta(i,s)$, and all $s\geq0$.

---

[25]   Note that condition (6.4.11) ensures that each cycle structure is left with probability 1 by some point in time $\tilde{t}\in\mathbb{R}_+$, whereas assumption $\overline{A3}$ stipulates that each cycle structure is left with positive probability at any time.

$F_{ij}^{\delta(i,s)}(s,\theta)=0$ says that, with probability 1, the duration of activity $<i,j>$ in the STEOR network N_δ is at least θ. As already mentioned in section 3.6, by an appropriate topological ordering of the nodes of the network (compare (1.1.1)), it can always be guaranteed that there is no arc $<i,j>$ with $i{\geq}j$ outside cycles and that there are only few arcs of that kind within cycles.

If assumption $\overline{A0}$ is satisfied, the value–iteration method or policy–improvement technique need not be used. Instead, Bellman's equation (6.3.3) (with "min" instead of "inf" and $\displaystyle\int_{\mathbb{R}_+} \dots$ replaced by $\displaystyle\int_0^{\overline{t}-s} \dots$) can be evaluated directly by the following recursion in state space. By assumption $\overline{A0}$ we have for $i{\geq}j$

$$\int_0^{\overline{t}-s} H^*(j,s+t)F_{ij}^{\delta(i,s)}(s,dt) = \begin{cases} \displaystyle\int_\theta^{\overline{t}-s} H^*(j,s+t)F_{ij}^{\delta(i,s)}(s,dt) & \text{for } s<\overline{t}-\theta \\[2em] 0 & \text{for } s\geq\overline{t}-\theta \end{cases}$$

Substituting into (6.3.3) yields for $(i,s)\in V\times[0,\overline{t}]$

$$(6.4.12) \quad H^*(i,s) = \min_{\delta(i,s)\in\Delta(i,s)} \left\{ h^{\delta(i,s)}(i,s) + \right.$$

$$+ \sum_{\substack{j\in S(i) \\ j>i}} p_{ij}^{\delta(i,s)}(s) \int_0^{\overline{t}-s} H^*(j,s+t)F_{ij}^{\delta(i,s)}(s,dt)$$

$$+ \begin{cases} \displaystyle\sum_{\substack{j\in S(i) \\ j\leq i}} p_{ij}^{\delta(i,s)}(s) \int_0^{\overline{t}-s} H^*(j,s+t)F_{ij}^{\delta(i,s)}(s,dt) & \text{for } s<\overline{t}-\theta \\[2em] 0 & \text{for } s\geq\overline{t}-\theta \end{cases}$$

Note that $H^*(i,\bar{t})=0$ for all $i \in V$ (because there is no terminal project cost) and $H^*(i,s)=0$ for each sink i and all $s \geq 0$. If the network is topologically ordered, node n is always a sink.

Now suppose that we want to evaluate equation (6.4.12) for $\bar{t}-\theta \leq s < \bar{t}$. To determine $H^*(i,s)$ for fixed $i \in V$ and $s \in [\bar{t}-\theta,\bar{t})$, we only need $H^*(j,t)$ for $\bar{t}-\theta \leq t < \bar{t}$ and $j>i$. Thus, it is expedient to evaluate (6.4.12) successively for $i=n-1,n-2,\ldots,1$ (we assume that the network is topologically ordered and thus $H^*(n,s)=0$ for all $s \geq 0$). In general, we suppose that $\bar{t}=\bar{k}\theta$ for some $\bar{k} \in \mathbb{N}$ and proceed as follows:

(6.4.13) **Algorithm.**

Step 1. Set $k:=\bar{k}$ and $H^*(n,s):=0$ for $s \geq 0$.

Step 2. Compute $H^*(i,s)$ for $(k-1)\theta \leq s < k\theta$ by means of (6.4.12) successively for $i=n-1,n-2,\ldots,1$

Step 3. If $k=1$, terminate;
         otherwise set $k:=k-1$ and to to step 2.   ∎

Chapter 7 Cost and Time Minimization for Decision Project Networks

In chapter 6 we have seen that the problem of minimizing the (expected) cost of a project described by a GERT network can be reduced to a stochastic dynamic programming problem. That approach, however, generally requires a great computational effort. In addition, it is only applicable if the GERT network represents an admissible EOR network (which includes the stipulation that the independence and Markov properties from assumptions A2b and A2d are satisfied).

In what follows, we will therefore present another approach to cost minimization for projects with stochastic evolution structure, which can also be applied to time minimization of projects. We start from a very simple acyclic deterministic project network which has a "GERT–like node logic" and is called a **decision project network**. The word "decision" indicates that we decide on which activities are to be carried out so as to minimize a certain objective function (for example, the total cost of the project). This leads to a combinatorial optimization problem. A special case of that concept called **decision CPM**, where there are only two types of nodes AND and OR, was presented by Crowston and Thompson (1967). Hindelang and Muth (1979) developed a dynamic programming algorithm for decision CPM.

To take into account a stochastic evolution structure of projects within the framework of decision project networks, we will introduce randomized actions and consider multiple successive executions of projects. It can then be shown that the minimum expected cost of a sequence of project executions until the project is completed successfully satisfies an optimality equation. By exploiting that optimality equation, a cost–minimal policy can be determined by a policy–iteration technique in a finite number of steps.

Finally, we will briefly deal with time minimization. It turns out that, in general, time minimization for decision networks is more complicated than cost minimization. Most of the following material on cost and time minimization can be found in Neumann (1984a) and Siedersleben (1982).

7.1 Decision Project Networks

Let N be an acyclic project network with sources and sinks (using activity–on–arc representation) whose node set is again designated by V and whose arc set is E. The arc weights of N will be specified later on. We specifically assume that N has exactly one source, which is denoted by r and corresponds to the beginning of the project in question. One of the sinks of N is supposed to represent the successful completion of the corresponding project and is denoted by s. The remaining sinks, if any, may represent different kinds of unsuccessful discontinuation of the project.

(7.1.1)   **Definition.**

An acyclic project network N with only one source and with sinks is called a **decision project network** or briefly **decision network** if each node i of N is assigned an **entrance characteristic** $\chi_i^- \in \{0, 1, \ldots, |\mathcal{P}(i)|\}$ and an **exit characteristic** $\chi_i^+ \in \{0, 1, \ldots, |\mathcal{S}(i)|\}$. The entrance and exit characteristics, which form the so–called **node logic**, have the following meaning:

(a) Node i is activated as soon as χ_i^- incoming activities have been terminated.

(b) As soon as node i has been activated, at most χ_i^+ outgoing activities are begun. If node i is not activated, no outgoing activity is carried out.

Sometimes it is expedient to replace "at most" in (7.1.1b) by "exactly". Note that conditions (7.1.1a) and (7.1.1b) imply that each activity is begun at its earliest possible start time.

For the source r we put

$$\chi_r^- := 0$$

that is, the source is always activated. Moreover,

$$\chi_i^+ := 0 \text{ for } i \in S$$

where S is again the sink set of N.

(7.1.2) Remarks.

(a) If $\chi_i^-=1$, node i has an IOR entrance, and if $\chi_i^-=|\mathcal{P}(i)|$, node i has an AND entrance. If "at most" is replaced by "exactly" in (7.1.1b), $\chi_i^+=1$ corresponds to a stochastic exit and $\chi_i^+=|\mathcal{S}(i)|$ corresponds to a deterministic exit.

(b) If a given decision network N has source set R with $|R|>1$ and, say, the set of sources $R'\subseteq R, R'\neq\emptyset$, is activated at the beginning of the project, we can formally transform N to a **corresponding one–source network** as follows (compare (1.3.4)): We introduce a new single source r_0 and, for each $i\in R$, the auxiliary arc $<r_0,i>$. Moreover, we set $\chi_{r_0}^- := \chi_{r_0}^+ :=0$ and

$$\chi_i^- := \begin{cases} 0 \text{ for } i\in R' \\ 1 \text{ for } i\in R\backslash R' \end{cases}$$

Thus, the assumption that a decision network has only one source does not mean any loss of generality. Note that assumption A1 for GERT networks is not required for decision networks with several sources. Obviously, none of the remaining assumptions A2 to A6 that are related to the stochastic structure and cycles of GERT networks is necessary, either.

Next, we introduce the **arc variables**

$$w_{ij} := \begin{cases} 1 \text{ if } <i,j> \text{ is carried out} \\ 0, \text{ otherwise} \end{cases} \quad (<i,j>\in E)$$

and **node variables**

$$u_i := \begin{cases} 1 \text{ if } i \text{ is activated} \\ 0, \text{ otherwise} \end{cases} \quad (i\in V)$$

where

$$u_r := 1$$

that is, source r is always activated. Then the **node–logic conditions** (7.1.1a) and (7.1.1b) can be rewritten as

$$(7.1.3) \qquad \left.\begin{array}{l} \displaystyle\sum_{k\in P(i)} w_{ki} \geq \overline{x}_i u_i \\[2ex] \displaystyle\sum_{k\in P(i)} w_{ki} < \overline{x}_i + M_i u_i \quad \text{where } M_i > |P(i)| - \overline{x}_i \end{array}\right\} \quad (i\in V\backslash\{r\})$$

$$(7.1.4)$$

$$(7.1.5) \qquad \sum_{j\in S(i)} w_{ij} \leq x_i^+ u_i \quad (i\in V\backslash S)$$

Putting $u_i:=1$ in (7.1.3) means that the activation of node i implies the execution of at least $\overline{x}_i$ incoming activities. Setting $u_i:=0$ in (7.1.4) says that if node i is not activated, then less than $\overline{x}_i$ incoming activities are carried out. Thus, both inequalities (7.1.3) and (7.1.4) together correspond to (7.1.1a). Putting $u_i:=1$ and $u_i:=0$ in (7.1.5) provides (7.1.1b).

(7.1.6) Since a decision network is acyclic, each activity of the corresponding project is either carried out exactly once or not executed at all. Thus, each **project realization** or **network realization** can be identified with the set of the activities carried out or with a function $w:E\to\{0,1\}$ whose function values are given by

$$w(\langle i,j\rangle)=:w_{ij} = \begin{cases} 1 \text{ if } \langle i,j\rangle \text{ is carried out} \\ 0, \text{ otherwise} \end{cases} \quad (\langle i,j\rangle\in E)$$

On the other hand, once a project realization w is given, the arc variables w_{ij} ($\langle i,j\rangle\in E$) and node variables u_i ($i\in V$) are specified. We speak of an **admissible project realization** w if w satisfies the node–logic conditions. Then

$$\mathcal{E}:=\{w:E\to\{0,1\}\,|\,w_{ij} \text{ satisfies (7.1.3), (7.1.4) and (7.1.5), } \langle i,j\rangle\in E\}$$

is the set of all admissible project realizations. Fig. 7.1.1 shows a decision network where an admissible project realization that activates sink $s=6$ is illustrated by darker arrows.

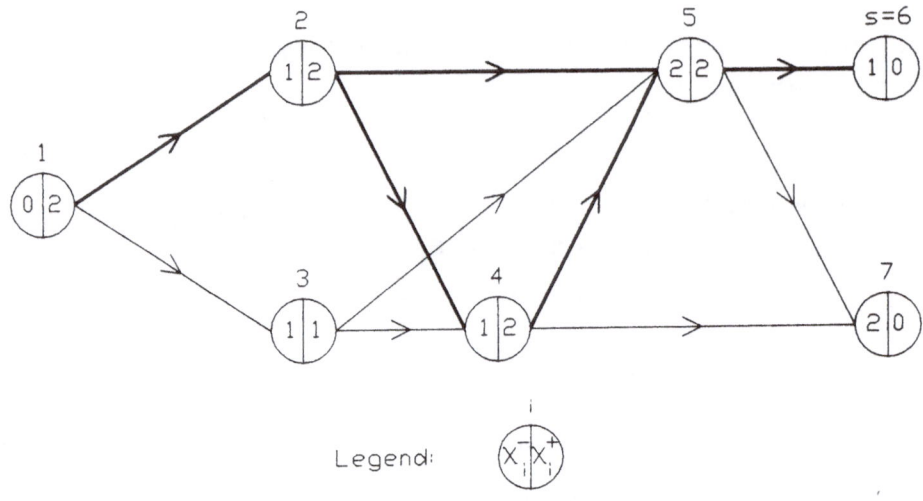

Figure 7.1.1

7.2 Cost Minimization

Now we assign a weight $c_{ij} \in \mathbb{R}_+$ to each arc $\langle i,j \rangle$ of the decision network N, which represents the cost arising when activity $\langle i,j \rangle$ is carried out. We then want to select a **successful project realization** $w \in \mathcal{E}$ with minimum cost, where "successful" means that the project realization results in the activation of sink s, which represents the successful completion of the project. This leads to the optimization problem

(7.2.1) $\qquad \left\{ \begin{array}{l} \text{Minimize} \quad c_w \\ \text{subject to} \ w \in \mathcal{E} \\ \qquad\qquad w \text{ activates s} \end{array} \right.$

where

(7.2.2) $\qquad c_w = \displaystyle\sum_{\langle i,j \rangle \in E} c_{ij} w_{ij}$

is the cost of project realization w. Optimization problem (7.2.1) suggests to consider a project realization w to be an **action** taken by a decision maker.

Before dealing with solving the optimization problem (7.2.1), we take a look at the recognition problem

(R)   Is there a $w \in \mathcal{E}$ that activates s?

In other words, (R) asks whether there is a feasible solution to optimization problem (7.2.1).

It can be shown that the decision problem (R) is in general NP–complete (for instance, there is a polynomial transformation from the non–tautology problem to problem (R), cf. Siedersleben (1982)). Hence, problem (7.2.1) is generally NP–hard. In the special cases where $\chi_i^-=1$ for all $i \in V \setminus \{r\}$ or $\chi_i^+=|S(i)|$ for all $i \in V \setminus S$ (for example, in acyclic EOR networks we have $\chi_i^-=1$ for $i \in V \setminus \{r\}$ and in CPM networks $\chi_i^+=|S(i)|$ for $i \in V \setminus S$), there is always a successful project realization.

The reason for the difficulty in solving problems (R) and (7.2.1) is the following: We know that source r and sink s have to be activated to get a successful project realization. But we do not know which additional nodes of the network are to be activated in order to activate sink s. If we, say in problem (7.2.1), prescribe the set of all nodes which are activated in project realization w, then this modified problem is related to the original problem (7.2.1) in the same manner as the minimal spanning tree problem is related to the Steiner network problem (which calls for spanning some specified nodes of a weighted graph by a tree with minimum weight). It is well–known that the minimal spanning tree problem can be solved in polynomial time but not the Steiner network problem (for the minimal spanning tree and Steiner network problems we refer to Lawler (1976), sections 7.2, 7.10, and 7.12).

By means of the arc variables w_{ij} and node variables u_i, the cost minimization problem (7.2.1) can be rewritten in the detailed form

(7.2.3)

$$
\begin{cases}
\text{Minimize} & \sum_{\langle i,j \rangle \in E} c_{ij} w_{ij} \\
\text{subject to} & \sum_{k \in P(i)} w_{ki} \geq \chi_i^- u_i \quad (i \in V \setminus \{r\}) \\
& \sum_{j \in S(i)} w_{ij} \leq \chi_i^+ u_i \quad (i \in V \setminus S) \\
& w_{ij} \in \{0,1\} \quad (\langle i,j \rangle \in E) \\
& u_i \in \{0,1\} \quad (i \in V \setminus \{s\}) \\
& u_s = 1
\end{cases}
$$

The first two constraints of problem (7.2.3) represent the node–logic conditions (7.1.3) and (7.1.5), whereas the remaining constraints say that the node and arc variables are binary variables and that sink s has to be activated. Since all activity costs are nonnegative and the total project cost is to be minimized, the execution of x_i^- activities leading into node i only makes sense if it implies the activation of node i. Thus, inequality (7.1.4), which ensures the latter fact, can be omitted.

To solve problem (7.2.3) we use a **branch–and–bound algorithm,** which is briefly sketched in what follows (for details of branch–and–bound methods we refer to Papadimitriou and Steiglitz (1982), chapter 18, and Nemhauser and Wolsey (1988), sections II.4.1 and II.4.2). Let A be the set of those nodes of the underlying decision network N which are activated and let $\bar{A}$ be the set of those nodes which are not activated in project realization w. Set A contains at least source r and sink s. Those nodes that neither belong to A nor to $\bar{A}$ may or may not be activated and are called **free nodes.**

Minimization problem (7.2.3) with the additional constraint

$$u_i = \begin{cases} 1 \text{ for } i \in A \\ 0 \text{ for } i \in \bar{A} \end{cases}$$

is designated by $(A,\bar{A})$ in what follows. Since given a project realization w, the corresponding node variables are uniquely specified, we briefly speak of a "solution w" to problem $(A,\bar{A})$ if the respective arc variables w_{ij} ($<i,j>\in E$) and the corresponding node variables u_i ($i \in V$) together represent a solution to $(A,\bar{A})$.

Problem (7.2.3) coincides with problem $(\{r,s\},\emptyset)$ and corresponds to the root of the branching tree associated with the branch–and–bound algorithm. For simplicity, the node in the branching tree corresponding to minimization problem $(A,\bar{A})$ is also denoted by $(A,\bar{A})$. If node $(A,\bar{A})$ is no leaf (that is, a sink) of the branching tree, it has the two successors $(A\cup\{k\},\bar{A})$ and $(A,\bar{A}\cup\{k\})$ where k is some free node of N.

In the course of the branch–and–bound algorithm, the best feasible solution to problem

(7.2.3) obtained so far is denoted by w^+. The algorithm uses two bounds. B is an upper bound on the minimum value of the objective function of problem (7.2.3). At the

beginning of the algorithm we may set $B := \sum_{<i,j>\in E} c_{ij}$. $b(A,\overline{A})$ is a lower bound on the

minimum value of the objective function of problem $(A,\overline{A})$. If $\overline{A}=V\backslash A$ (that is, there are

no free nodes in N), $b(A,\overline{A})$ equals the minimum value of the objective function of

problem $(A,\overline{A})$. An appropriate bound function b will be proposed later on.

A leaf of the branching tree is called **active** if it has not been examined yet. At the beginning of the algorithm, the root $(\{r,s\},\emptyset)$ is the only active node. An active node

is examined as follows. If $b(A,\overline{A}) \geq B$, then node $(A,\overline{A})$ can be removed from the

branching tree. When $b(A,\overline{A})<B$ and $\overline{A}=V\backslash A$, we put $w^+:=w$ and $B:=b(A,\overline{A})$ where w is a

solution to problem $(A,\overline{A})$. If $b(A,\overline{A})<B$ and there is at least one free node in N, then the

two successors $(A\cup\{k\},\overline{A})$ and $(A,\overline{A}\cup\{k\})$ of node $(A,\overline{A})$ are generated where k is any

free node of N ("branching from node $(A,\overline{A})$"). $(A\cup\{k\},\overline{A})$ and $(A,\overline{A}\cup\{k\})$ represent new active nodes of the branching tree.

The active nodes are put on a list L. If L contains more than one element, we have to decide which node from L should be examined next. There are three basic rules that choose a node from L. Depth–first search or LIFO (last in, first out) means that L is implemented as a stack. In breadth–first search or FIFO (first in, first out), L is implemented as a queue. Thirdly, L can be implemented as a heap such that the node

$(A,\overline{A})$ from L to be examined next has least $b(A,\overline{A})$. For the data structures stack,

queue and heap we refer to Aho et al. (1983), sections 2.3, 2.4 and 4.11.

The branch–and–bound algorithm terminates when $L=\emptyset$. Then the best feasible solution

w^+ obtained so far is optimal.

An appropriate bound $b(A,\overline{A})$ is given by the minimum value of the objective function of optimization problem

$$
(7.2.4) \quad
\begin{cases}
\text{Minimize} & \sum\limits_{<i,j>\in E} c_{ij} w_{ij} \\[2mm]
\text{subject to} & \sum\limits_{k\in P(i)} w_{ki} = x_i^- \qquad (i\in A\setminus\{r\}) \\[2mm]
& \sum\limits_{j\in S(i)} w_{ij} \leq
\begin{cases}
0 & \text{for } i\in \overline{A}\setminus S \\
x_i^+ & \text{for } i\in (V\setminus\overline{A})\setminus S
\end{cases} \\[4mm]
& w_{ij}\in\{0,1\} \qquad (<i,j>\in E)
\end{cases}
$$

Problem (7.2.4) results from problem (7.2.3) to be solved by specifying the node variables such that $u_i=1$ for $i\in A$ (and $\geq$ is replaced by $=$) in the first constraint as well as $u_i=0$ for $i\in\overline{A}$ and $u_i=1$ for $i\in V\setminus\overline{A}$ in the second constraint. Note that (7.2.4) coincides with problem $(A,\overline{A})$ if $\overline{A}=V\setminus A$. Problem (7.2.4) is equivalent to finding an integral minimum cost flow of the given value $v=\sum\limits_{j\in A} x_j^-$ from source a to sink b in the network $N_{A,\overline{A}}$ shown in Fig. 7.2.1 (for minimum cost flows we refer to Lawler (1976), section 4.7, or Neumann (1987a), section 6.4.5).

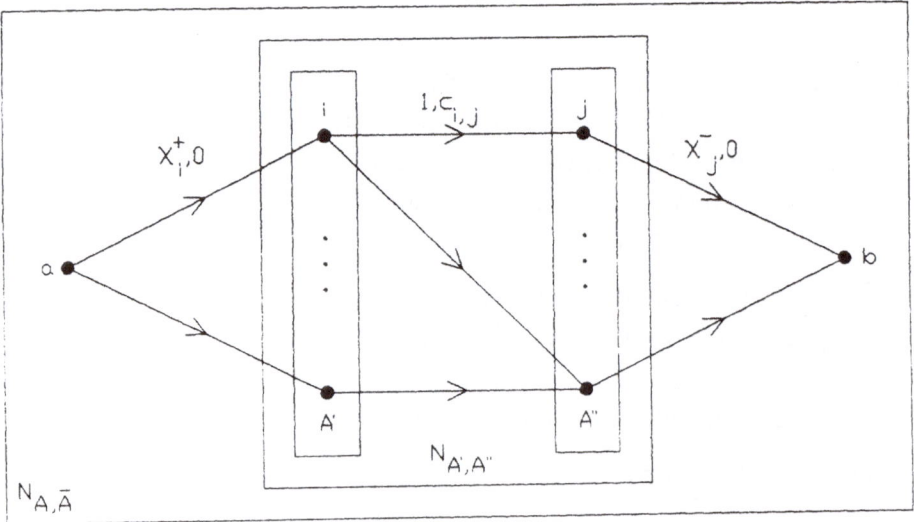

Figure 7.2.1

Network $N_{A,\bar{A}}$ contains all arcs $<i,j>$ of N where $i\in(V\backslash\bar{A})\backslash S$ (that is, node i may be activated) and $j\in A\backslash\{r\}$ (that is, node j must be activated). Let A' and A" be the sets of the latter nodes i and j, respectively, In addition to the nodes from A' and A" and the above arcs, network $N_{A,\bar{A}}$ contains source a and sink b as well as the arcs $<a,i>$ $(i\in A')$ and $<j,b>$ $(j\in A")$. Each arc $<\alpha,\beta>$ of $N_{A,\bar{A}}$ is labelled $\gamma_{\alpha\beta}, c_{\alpha\beta}$ in Fig. 7.2.1 where $\gamma_{\alpha\beta}$ is the capacity and $c_{\alpha\beta}$ is the cost of unit flow for arc $<\alpha,\beta>$.

Problem (7.2.4) can also be considered a generalized weighted matching problem in the network $N_{A',A"}$ of Fig. 7.2.1 with weights $c_{\alpha\beta}$ (for the "standard" weighted matching problem it holds that $\chi_i^+=1$ $(i\in A')$ and $\chi_j^-=1$ $(j\in A")$, cf. Lawler (1976), sections 5.1 and 5.2). Problem (7.2.4) can be solved by well–known min–cost flow methods, for example, a successive shortest path algorithm or a so–called scaling technique (cf. Nemhauser et al. (1989), section IV.5) or the network simplex method (cf. Chvatal (1983), chapters 19 and 21), or by an appropriately modified algorithm for the weighted matching problem (compare Lawler (1976), section 5.8). For a detailed version of the branch–and–bound algorithm sketched above we refer to Siedersleben (1982). The latter reference also shows how to find tighter lower bounds $b(A,\bar{A})$ by adding some further constraints to problem (7.2.4), which results in a min–cost flow problem with multipliers.

The branch–and–bound method for solving problem (7.2.3) is summarized as

(7.2.5) Algorithm.

Step 1. Set $B:=\sum\limits_{<i,j>\in E} c_{ij}$ and $L:=\{(\{r,s\},\emptyset)\}$

Step 2. If $L=\emptyset$, then terminate (w^+ is optimal; if no w^+ has been found, there is no feasible solution to (7.2.3));
otherwise choose a node $(A,\bar{A})$ from L (by using one of the three basic rules stated above), delete $(A,\bar{A})$ from L and compute $b(A,\bar{A})$ by solving problem (7.2.4)

Step 3. If $b(A,\bar{A})<B$ and there is no free node in N (that is, $\bar{A}=V\backslash A$), set $w^+:=w$ and $B:=b(A,\bar{A})$ where w is a solution to (7.2.4)
If $b(A,\bar{A})<B$ and there are free nodes in N, choose any free node k in N and insert $(A\cup\{k\},\bar{A})$ and $(A,\bar{A}\cup\{k\})$ in L
Go to step 2   ∎

7.3 Randomized Actions

The concept of the decision network introduced in section 7.1 leads to a deterministic model. The deterministic actions or project realizations include only the two cases that an activity is either carried out with certainty or not carried out at all. To cope with uncertainties during the evolution of a project, we introduce randomized actions, which cover the case that an activity is executed with a probability between 0 and 1. In addition randomized actions may result in less (expected) cost than deterministic actions.

(7.3.1)   **Definition.**

A **randomized action** is a mapping $\pi:\mathcal{E}\to[0,1]$ with $\sum_{w\in\mathcal{E}}\pi(w)=1$. The (admissible) project realizations $w\in\mathcal{E}$ are also called (admissible) **deterministic actions.**

The concept of a randomized action corresponds to the concept of a mixed strategy in game theory, and the deterministic actions correspond to the pure strategies (for the elements of game theory we refer to Chvatal (1983), chapter 15). The set Π of all randomized actions is isomorphic to (and can thus be identified with) the convex hull of the set $\mathcal{E}$ of the admissible deterministic actions, which represents a convex polyhedron.

(7.3.2) In (1.2.1) we have defined a project execution to be a performance of the basic random experiment and a project realization to be an outcome of that experiment, that is, an element ω of the underlying sample space Ω. Now suppose that a probability space $(\Omega,\mathcal{A},P_\pi)$ is given for each $\pi\in\Pi$ and there is a one–to–one mapping of Ω onto $\mathcal{E}$, $W:\Omega\to\mathcal{E}$. Thus, the elements $\omega\in\Omega$ can be identified with the corresponding elements $W(\omega)\in\mathcal{E}$. The random variable W can be considered a **project execution** (or **network execution**), and the function values $W(\omega),w\in\Omega$, represent project realizations in the sense of (7.1.6). $\pi(w)$, also denoted by π_w in what follows, may then be viewed as the probability that project realization $w\in\mathcal{E}$ occurs if randomized action π is taken:

$$\pi(w) = P_\pi\{\omega\in\Omega\,|\,W(\omega)=w\}:=\pi_w$$

Next, we consider some further probabilities:

The probability that activity $<i,j>$ is carried out if randomized action π is chosen:

$$(7.3.3) \qquad \rho_{ij}(\pi) := P_{\pi}(<i,j> \text{ is carried out}) = \sum_{w \in \mathcal{E}} w_{ij} \pi_w$$

The probability that node i is activated if randomized action π is taken:

$$(7.3.4) \qquad q_i(\pi) := P_{\pi}(i \text{ is activated}) = \sum_{w \in \mathcal{E}} u_i \pi_w$$

The conditional probability that activity $<i,j>$ is carried out given that the initial event i of that activity has occurred if randomized action π is selected:

$$(7.3.5) \qquad p_{ij}(\pi) := P_{\pi}(<i,j> \text{ is carried out} \mid i \text{ has occurred})$$

$$= \begin{cases} \dfrac{\rho_{ij}(\pi)}{q_i(\pi)} & \text{for } q_i(\pi) > 0 \\ \\ 0 & \text{, otherwise} \end{cases}$$

We see that all those probabilities can be expressed in terms of the arc and node variables and the probabilities π_w. Recall that once a project realization w is given, the arc variables w_{ij} ($<i,j> \in E$) and node variables u_i ($i \in V$) are specified (cf. (7.1.6)).

At last we consider the expected cost $c(\pi)$ incurred by randomized action π. Let $\mathbb{E}_{\pi}$ denote the expected value with respect to the probability measure P_{π} and let c_W be the cost of project execution W. Then by (7.2.2) and (7.3.3)

$$(7.3.6) \qquad c(\pi) := \mathbb{E}_{\pi}(c_W) = \sum_{w \in \mathcal{E}} c_w \pi_w = \sum_{<i,j> \in E} c_{ij} \rho_{ij}(\pi)$$

which is finite for every $\pi \in \Pi$. We seek a randomized action that minimizes this expected project cost.

We assume that for each activity $<i,j>$ there are a smallest probability λ_{ij} and a greatest probability κ_{ij} that $<i,j>$ is carried out given that project event i has occurred:

(7.3.7) $0 \leq \lambda_{ij} \leq p_{ij}(\pi) \leq \kappa_{ij} \leq 1$ if $q_i(\pi) > 0$ ($<i,j> \in E$)

Moreover, we only take into account randomized actions π for which the probability $q(\pi) := q_s(\pi)$ that sink s is activated (that is, the probability of a successful completion of the project) is not less than a prescribed positive bound ϵ:

(7.3.8) $q(\pi) := q_s(\pi) \geq \epsilon$ where $0 < \epsilon \leq 1$

(7.3.9) A randomized action satisfying (7.3.7) and (7.3.8) is said to be **admissible**, and

$$\Pi^{\times} := \{\pi \in \Pi \mid \pi \text{ satisfies } (7.3.7) \text{ and } (7.3.8)\}$$

is the set of the admissible randomized actions. $\Pi^{\times}$ is again isomorphic to a convex polyhedron. In what follows we assume that $\Pi^{\times} \neq \emptyset$.

The cost minimization problem when randomized actions are present becomes

(7.3.10) $\left[\begin{array}{l} \text{Minimize} \quad c(\pi) \\ \\ \text{subject to} \quad \pi \in \Pi^{\times} \end{array} \right.$

The following theorem shows under which conditions a deterministic action is an optimal solution to (7.3.10).

(7.3.11)   Theorem.
          There is a deterministic optimal action if all bounds are integral [26].

Proof.

Since $\Pi^{\times}$ is nonvoid and compact, and $c(\cdot)$ is continuous, (7.3.10) has an optimal solution π^*. Let $w^+ \in \mathcal{E}$ be such that $\pi^*_{w^+} > 0$ and

(7.3.12) $c_{w^+} = \min\{c_w \mid \pi^*_w > 0, w \in \mathcal{L}\}$

---

[26]   That is, λ_{ij} and κ_{ij} equal 0 or 1 for each $<i,j> \in E$ and $\epsilon=1$.

Let π^+ be that element from Π which corresponds to deterministic action w^+, that is,

$$\pi_w^+ := \begin{cases} 1 \text{ for } w=w^+ \\ 0, \text{ otherwise} \end{cases} \quad (w \in \mathcal{E})$$

Since all bounds are integral, we have $\pi^+ \in \Pi^\times$ and by (7.3.12)

$$c(\pi^+) = c_{w^+} \leq c(\pi^*)$$

∎

Owing to theorem (7.3.11) a decrease in the (expected) project cost by using admissible randomized actions instead of admissible deterministic actions is only possible if some bounds are not integral.

Using (7.3.3) to (7.3.8) problem (7.3.10) can be rewritten as

(7.3.13)

$$\begin{cases} \text{Minimize} & \sum_{w \in \mathcal{E}} c_w \pi_w \\ \text{subject to} & \left. \begin{array}{l} \sum_{w \in \mathcal{E}} w_{ij} \pi_w - \lambda_{ij} \sum_{w \in \mathcal{E}} u_i \pi_w \geq 0 \\ \kappa_{ij} \sum_{w \in \mathcal{E}} u_i \pi_w - \sum_{w \in \mathcal{E}} w_{ij} \pi_w \geq 0 \end{array} \right\} \quad (<i,j> \in E) \\ & \sum_{w \in \mathcal{E}} u_s \pi_w \geq \epsilon \\ & \sum_{w \in \mathcal{E}} \pi_w = 1 \\ & \pi_w \geq 0 \quad (w \in \mathcal{E}) \end{cases}$$

The first two constraints of problem (7.3.13) correspond to condition (7.3.7) and the third constraint corresponds to (7.3.8). The last two constraints say that the π_w ($w \in \mathcal{E}$) represent probabilities. Note that, for $<i,j> \in E$, the first constraint is to be dropped if $\lambda_{ij}=0$ and the second is dropped if $\kappa_{ij}=1$.

Problem (7.3.13) represents a linear programming problem whose variables are the probabilities π_w ($w \in \mathcal{E}$). Let m_1 be the number of positive λ_{ij} and m_2 be the number of the κ_{ij} that are less than 1. Since the number of functional constraints of problem

(7.3.13), m_1+m_2+2, is generally small in comparison with the number $|\mathcal{E}|$ of variables, the revised simplex method is recommended to solve (7.3.13) (for the revised simplex method we refer to Chvatal (1983), chapter 7). This method only needs the inverse of the current $(m_1+m_2+2) \times (m_1+m_2+2)$ basis matrix in each iteration step. Furthermore, each iteration step calls for selecting a $w \in \mathcal{E}$ such that the "reduced cost coefficient" of the corresponding nonbasic variable π_w, say $\bar{c}_w$, is negative. In other words, each iteration step of the simplex method requires the computation of a feasible solution to a problem of type (7.2.1) (without the last constraint "w activates s") with negative value of the objective function (to check whether the optimality criterion $\min_{w \in \mathcal{E}} \bar{c}_w \geq 0$ is satisfied, an optimal solution to a problem of type (7.2.1) has to be found at least in the last iteration step).

7.4 Multiple Executions of Projects

Since a decision network is assumed to be acyclic, each activity can be carried out at most once during a single project realization. In practice, however, multiple executions of activities or of parts of the project or of the entire project can occur if parts of the project or the whole project are terminated unsuccessfully. To this end, we consider a sequence of randomized actions (where each randomized action implies a project execution) until the project is completed successfully (that is, sink s is activated). In other words, we have to solve some kind of **stopping problem.**

Let $W_1, W_2, \ldots$ be a sequence of project executions (where $W_\nu : \Omega \to \mathcal{E}$ for $\nu \in \mathbb{N}$). We assume that the family of random variables $(W_\nu)_{\nu \in \mathbb{N}}$ is stochastically independent. Let

$$Z_\nu := \begin{cases} 1 & \text{if sink } s \text{ is activated in project execution } W_\nu \\ 0, & \text{otherwise} \end{cases} \quad (\nu \in \mathbb{N})$$

and let

$$M := \inf\{\nu \in \mathbb{N} \mid Z_\nu = 1\}$$

be the number of project executions up to (and including) the successful completion of the project.

(7.4.1) A sequence $\psi := (\pi^1, \pi^2, \ldots)$ of admissible randomized actions (that is, $\pi^\nu \in \Pi^\times$ for $\nu \in \mathbb{N}$) is called an **admissible policy**. The set of all admissible policies is denoted by Ψ. The expected total cost up to the successful completion of the project, if policy ψ is used, is

$$(7.4.2) \qquad C(\psi) := \mathbb{E}_\psi \left[\sum_{\nu=1}^{M} c_{W_\nu} \right]$$

where the subscript ψ of $\mathbb{E}_\psi$ refers to the probability measure P_ψ related to policy ψ. We look for an admissible policy that minimizes the expected total cost. This gives the minimization problem

$$(7.4.3) \qquad \begin{cases} \text{Minimize} \quad C(\psi) \\ \text{subject to } \psi \in \Psi \end{cases}$$

To solve problem (7.4.3) we need some theorems. Let

$$C^* := \inf_{\psi \in \Psi} C(\psi)$$

Recall that $q(\pi)$ is the probability of a successful project execution and $c(\pi)$ is the expected project cost if randomized action π is taken. Then we have

(7.4.4)   **Theorem.**

For all $\psi \in \Psi$ it holds that

(a) $\min_{\pi \in \Pi^\times} [c(\pi) - q(\pi) C(\psi)] \leq 0$

and C^* satisfies the **optimality equation**

(b) $\min_{\pi \in \Pi^\times} [c(\pi) - q(\pi) C^*] = 0$

Proof.

Let $\psi := (\pi^1, \pi^2, \pi^3, \ldots) \in \Psi$ and $\hat{\psi} := (\pi^2, \pi^3, \ldots)$. Exploiting the independence of the random variables $W_1, W_2, \ldots$ we obtain by using (7.4.2)

$$C(\psi) = c(\pi^1) + [1 - q(\pi^1)] C(\hat{\psi})$$

and

(7.4.5) $\qquad C^* = \inf_{\psi \in \Psi} C(\psi) = \inf_{\substack{\psi \in \Psi \\ \pi^1 \in \Pi^\times}} \{ c(\pi^1) + [1 - q(\pi^1)] C(\hat{\psi}) \}$

$$= \inf_{\pi^1 \in \Pi^\times} \{ c(\pi^1) + [1 - q(\pi^1)] C^* \}$$

Since $\Pi^\times$ is nonempty and compact and since $c(\cdot)$ and $q(\cdot)$ are continuous, we may write min instead of inf in (7.4.5). Subtraction of C^* on the left hand side and right hand side of (7.4.5) provides the optimality equation (7.4.4b). Because of $q(\pi) \geq 0$ for all $\pi \in \Pi^\times$ and $C(\psi) \geq C^*$ for all $\psi \in \Psi$ we have

$$c(\pi) - q(\pi) C(\psi) \leq c(\pi) - q(\pi) C^* \text{ for all } \pi \in \Pi^\times \text{ and all } \psi \in \Psi$$

Using (7.4.4b) we then obtain (7.4.4a).

∎

(7.4.6) An admissible policy $\psi_\pi := (\pi, \pi, \ldots)$, $\pi \in \Pi^\times$, is called **stationary**. Exploiting again the independence of the random variables $W_1, W_2, \ldots$, it holds for the expected number of project executions until the successful completion of the project if stationary policy ψ_π is used that

$$\mathbb{E}_{\psi_\pi}(M) = 1 + [1 - q(\pi)] \mathbb{E}_{\psi_\pi}(M | Z_1 = 0) = 1 + [1 - q(\pi)] \mathbb{E}_{\psi_\pi}(M)$$

and thus

(7.4.7) $\mathbb{E}_{\psi_\pi}(M) = \frac{1}{q(\pi)}$

Since $q(\pi) \geq \epsilon > 0$ for all $\pi \in \Pi^\times$ (compare (7.3.8)), the expected number of project executions up to the successful completion of the project is bounded above by $\frac{1}{\epsilon}$ if stationary admissible policies are used. By (7.4.2) and Wald's equation (cf. Rohatgi (1976), section 14.3) we obtain

$$C(\psi_\pi) = \mathbb{E}_{\psi_\pi}\left[\sum_{\nu=1}^{M} c_{W_\nu}\right] = \mathbb{E}_{\psi_\pi}(M)\mathbb{E}_\pi(c_W)$$

Then (7.3.6) and (7.4.7) give

(7.4.8) $C(\psi_\pi) = \frac{c(\pi)}{q(\pi)}$ for all $\pi \in \Pi^\times$

Next, we state

(7.4.9)   Theorem.
            There exists a stationary optimal policy, that is, a policy ψ_{π^+} such that

$C(\psi_{\pi^+}) = C^*$.

Proof.

Let $\pi^+ \in \Pi^\times$ be such that in (7.4.4b) the minimum is attained for π^+. Then

$$c(\pi^+) - q(\pi^+)C^* = 0$$

and (7.4.8) yield $C(\psi_{\pi^+}) = C^*$.

∎

Owing to theorem (7.4.9) we may restrict ourselves to stationary policies when looking for an optimal solution to problem (7.4.3). By theorem (7.4.4), for each optimal policy the equality sign holds in (7.4.4a). The following theorem says that for stationary policies the converse is also true.

(7.4.10)   Theorem.

If $\hat{\pi} \in \Pi^\times$ satisfies the optimality equation

(a) $\min_{\pi \in \Pi^\times} [c(\pi) - q(\pi) C(\psi_{\hat{\pi}})] = 0$

then $\psi_{\hat{\pi}}$ is a (stationary) optimal policy.

Proof.
By (7.4.10a)

$$c(\pi) - q(\pi) C(\psi_{\hat{\pi}}) \geq 0 \quad \text{for all } \pi \in \Pi^\times$$

and thus by (7.4.8)

$$C(\psi_{\hat{\pi}}) \leq C(\psi_\pi) \quad \text{for all } \pi \in \Pi^\times$$

In particular, the latter inequality holds for $\pi = \pi^+$ where ψ_{π^+} is a stationary optimal policy, which exists by theorem (7.4.9). Hence, we have

$$C(\psi_{\hat{\pi}}) \leq C(\psi_{\pi^+}) = C^*$$

and thus $C(\psi_{\hat{\pi}}) = C^*$, that is, $\psi_{\hat{\pi}}$ is optimal.   ∎

Theorem (7.4.10) permits us to use the following **policy–iteration technique** for solving optimization problem (7.4.3). Start with any admissible stationary policy ψ_{π^0} ($\pi^0 \in \Pi^\times$) and put

$$C_0 := C(\psi_{\pi}0)$$

In iteration step $\nu \geq 1$ find $\pi^{\nu} \in \Pi^{\times}$ such that

(7.4.11) $c(\pi^{\nu}) - C_{\nu-1}q(\pi^{\nu}) = \min_{\pi \in \Pi^{\times}}[c(\pi) - C_{\nu-1}q(\pi)]$

where $C_{\nu-1} := C(\psi_{\pi}\nu-1)$.

Because of (7.4.4a) only the following two cases are possible:

Case 1. $\min_{\pi \in \Pi^{\times}}[c(\pi) - C_{\nu-1}q(\pi)] = 0$

Case 2. $\min_{\pi \in \Pi^{\times}}[c(\pi) - C_{\nu-1}q(\pi)] < 0$

In case 1, from theorem (7.4.10) it follows that policy $\psi_{\pi}\nu-1$ is optimal. In case 2, (7.4.11) and (7.4.8) provide

(7.4.12) $\min_{\pi \in \Pi^{\times}}[c(\pi) - C_{\nu-1}q(\pi)] = c(\pi^{\nu}) - C_{\nu-1}q(\pi^{\nu}) < 0 = c(\pi^{\nu}) - C_{\nu}q(\pi^{\nu})$

where $C_{\nu} := C(\psi_{\pi}\nu)$. (7.4.12) says that $C_{\nu} < C_{\nu-1}$, in other words, each iteration step provides a policy better than the preceding one until an optimal policy is attained.

(7.4.13)   **Algorithm (Policy iteration).**

Step 1.   Choose an arbitrary $\pi \in \Pi^{\times}$ and set $C := \dfrac{c(\pi)}{q(\pi)}$

Step 2.   Find $\pi^{+} \in \Pi^{\times}$ such that $c(\pi^{+}) - Cq(\pi^{+}) = \min_{\pi \in \Pi^{\times}}[c(\pi) - Cq(\pi)]$

Step 3.   If $c(\pi^{+}) - Cq(\pi^{+}) = 0$, then terminate ($\psi_{\pi^{+}}$ is optimal);

otherwise set $C := \dfrac{c(\pi^{+})}{q(\pi^{+})}$ and go to step 2

■

Each iteration step of the policy–iteration technique calls for solving a linear programming problem of the form

(7.4.14) $\quad \begin{cases} \text{Minimize} & c(\pi) - Cq(\pi) \\ \text{subject to} & \pi \in \Pi^{\times} \end{cases}$

where $C \in \mathbb{R}$ is known from the preceding iteration. Problem (7.4.14) is of type (7.3.10) and can be solved by means of the revised simplex method. Since the simplex method determines an optimal solution from the finitely many vertices of the convex polyhedron $\Pi^{\times}$ and, as seen above, a sequence of distinct vertices $\pi^{\nu} \in \Pi^{\times}$ is generated by the policy–improvement technique, a (stationary) optimal policy is attained in a finite number of iterations.

In Neumann (1984a) and Siedersleben (1982) it is shown that, besides the policy–iteration technique, a subgradient method can be used for solving problem (7.4.3).

In conclusion, we summarize the three models and corresponding cost minimization problems discussed in sections 7.2, 7.3 and 7.4.

The first model presented in section 7.2 is a deterministic one and permits activities to be carried out either exactly once or not at all during any project execution. The corresponding cost minimization problem

$$\begin{array}{ll} \text{Minimize} & c_w \\ \text{subject to} & w \in \mathcal{E} \\ & w \text{ activates } s \end{array}$$

whose feasible solutions are the "successful" admissible project realizations w can be solved by means of a branch–and–bound procedure.

The second model treated in section 7.3 permits activities to be carried out with a probability between 0 and 1. The feasible solutions to the corresponding optimization problem

$$\text{Minimize} \quad c(\pi)$$

$$\text{subject to} \quad \pi \in \Pi^{\times}$$

where $\Pi^{\times}$ is a convex polyhedron are the admissible randomized actions π. This optimization problem can be solved using the revised simplex method.

The last model discussed in this section deals with sequences of randomized actions and thus multiple project executions up to the successful complection of the project. The corresponding cost minimization problem

$$\text{Minimize} \quad C(\psi)$$

$$\text{subject to} \quad \psi \in \Psi$$

where the policies ψ are sequences of randomized actions can be solved by a policy–iteration technique in a finite number of steps.

7.5 Time Minimization

In this section we consider a decision network N where the weight of arc $<i,j>\in E$ is the duration $d_{ij} \in \mathbb{R}_+$ of the corresponding activity [27] instead of cost c_{ij} as before. We will only deal with deterministic actions or project realizations $w \in \mathcal{E}$. The models for randomized actions and for policies (sequences of randomized actions) are similar to the corresponding cost minimization models treated in sections 7.3 and 7.4.

Let d_w be the duration of project realization w, that is, the time needed for carrying out all activities $<i,j>$ with $w_{ij}=1$ (recall that because of (7.1.1a) and (7.1.1b) each activity is begun at its earliest possible start time given the project realization w in question). Then, instead of (7.2.1), we have the optimization problem

---

[27] We denote durations and later on times of activation by small letters (instead of capital letters as in the preceding chapters) because these quantities are now deterministic.

$$(7.5.1) \quad \left\{ \begin{array}{l} \text{Minimize} \quad d_w \\ \text{subject to } w \epsilon \mathcal{E} \\ \qquad\qquad w \text{ activates } s \end{array} \right.$$

Let

$$(7.5.2) \qquad \mathcal{E}^{\times} := \{w \epsilon \mathcal{E} \,|\, w \text{ activates } s\}$$

be the set of the successful project realizations. For $\mathcal{E}^{\times} \neq \emptyset$,

$$(7.5.3) \qquad d^* := \min_{w \epsilon \mathcal{E}^{\times}} d_w$$

is then the minimum value of the objective function of problem (7.5.1).

Suppose that, in accordance with convention (1.3.1), each project realization begins with the activation of source r at time 0. For $w \epsilon \mathcal{E}$, let t_j^w be the time of activation of node $j \epsilon V$ in project realization w, where $t_r^w = 0$ and $t_j^w := \infty$ if node j is not activated during project realization w. Obviously, for $j \epsilon V \setminus \{r\}$

$$(7.5.4) \qquad t_j^w = \min\{t \geqslant 0 \,|\, \text{there are } \chi_j^- \text{ different } i \epsilon P(j)$$
$$\text{such that } w_{ij} = 1 \text{ and } t_i^w + d_{ij} \leqslant t\}$$

where $\min \emptyset := \infty$. Furthermore, it holds that

$$d_w \geqslant t_s^w \quad \text{for all } w \epsilon \mathcal{E}^{\times}$$

We have $d_w > t_s^w$ if sink s is activated while some activities $\langle i, j \rangle$ with $w_{ij} = 1$ are still being carried out.

Let

$$t_j^e := \min_{w \epsilon \mathcal{E}} t_j^w$$

be the earliest possible time of activation of node j during any admissible project realization. Obviously,

(7.5.5) $$t_s^e = \min_{w \in \mathcal{E}^x} \max_{<i,j> \in E} \{(t_i^w + d_{ij}) w_{ij}\}$$

Let

(7.5.6) $$E_w := \{<i,j> \in E \mid w_{ij} = 1\}$$

be the support of function w, that is, the set of activities carried out during project realization w. Then we prove

(7.5.7)   **Theorem.**
The minimum duration of a successful project realization equals the earliest possible time of activation of sink s: $d^* = t_s^e$ for $\mathcal{E}^x \neq \emptyset$.

Proof.

Since $d_w \geq t_s^w$ for all $w \in \mathcal{E}^x$, we only have to show that $d^* \leq t_s^e$. Let $w^+ \in \mathcal{E}^x$ such that $t_s^{w^+} = t_s^e$, and let E' be the set of all activities $<i,j> \in E_{w^+}$ which are not terminated yet by time t_s^e. Recall that each project realization w is uniquely specified by its support E_w (cf. (7.5.6)). Since after the activation of a node k, at most χ_k^+ outgoing activities must be begun (cf. (7.1.1b)), $E_{w^+} \setminus E'$ represents the support of a "minimal" project realization $w^* \in \mathcal{E}^x$ with $d_{w^*} = t_s^{w^+}$. Therefore, $d^* = \min_{w \in \mathcal{E}^x} d_w \leq d_{w^*} = t_s^e$.

∎

Theorem (7.5.7) says that we can find the quantity d^* we are looking for by computing the earliest possible times of activation t_j^e $(j \in V)$. The latter problem, however, is in general NP–hard (recall that even the decision problem whether there is a successful project realization is generally NP–complete). Nevertheless, in the special case where $\chi_j^+ = |S(j)|$ for all $j \in V \setminus S$, the times t_j^e can be found by evaluating a simple optimality equation. First we prove

(7.5.8)   **Theorem.**

For $j \in V \setminus \{r\}$ it holds that

$$t_j^e \geq \bar{t}_j := \min\{t \geq 0 \mid \text{ there are } \chi_j^- \text{ different } i \in P(j) \text{ such that } t_i^e + d_{ij} \leq t\}$$

Proof.

Let $j \in V \setminus \{r\}$. Trivially, $t_j^e = \bar{t}_j = \infty$ precisely if there is no $w \in \mathcal{E}$ that activates node j. Now let $w \in \mathcal{E}$ such that $t_j^w = t_j^e$. For $i \in P(j)$, we have $t_i^w \geq t_i^e$ by definition of t_i^e. To activate node j at time t_j^e there have to be χ_j^- different $i \in P(j)$ which have been activated such that $t_i^w + d_{ij} \leq t_j^e$. Thus,

$$t_j^e = \min\{t \geq 0 \mid \text{ there are } \chi_j^- \text{ different } i \in P(j) \text{ such that } t_i^w + d_{ij} \leq t\} \geq \bar{t}_j \quad \blacksquare$$

The next theorem contains the optimality equation mentioned.

(7.5.9)   **Theorem.**

Let

(a) $\chi_i^+ = |\mathcal{S}(i)|$ for all $i \in V \setminus S$

Then

$$t_j^e = \bar{t}_j := \min\{t \geq 0 \mid \text{ there are } \chi_j^- \text{ different } i \in P(j) \text{ such that } t_i^e + d_{ij} \leq t\}$$

$$(j \in V \setminus \{r\})$$

Proof (cf. Siedersleben (1982)).

Let $j \in V \setminus \{r\}$. By (7.5.9a) there is a $w \in \mathcal{E}$ which activates node j. Because of theorem (7.5.8), we only have to show that $t_j^e \leq \bar{t}_j$. For $i \in P(j)$, let $w_i \in \mathcal{E}$ such that $t_i^{w_i} = t_i^e$. By (7.5.9a) the project realization w' with support

$$E_{w'} := \bigcup_{i \in P(j)} E_{w_i}$$

is admissible and activates each $i \in P(j)$ at time t_i^e. Let $\dot{E}$ be the set of those arcs $\langle i, j \rangle$

which furnish the χ_j^- smallest values $t_j^e + d_{ij}$, $i \in P(j)$ (there may be more than one such $\tilde{E}$). Then the project realization $\bar{w}$ with support $E_{\bar{w}} := E_w \cup \tilde{E}$, which is again admissible, activates node j at time $\bar{t}_j$ and we have $\bar{t}_j \geq t_j^e$.

∎

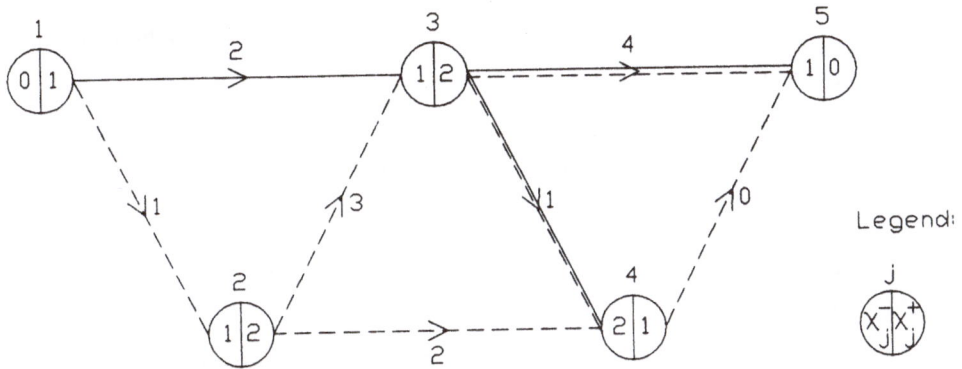

Figure 7.5.1

Node j	t_j^w	$t_j^{w'}$	t_j^e
1	0	0	0
2	∞	1	1
3	2	4	2
4	∞	5	5
5	6	5	5

Table 7.5.1

As an example, we consider the decision network with arc set E shown in Fig. 7.5.1, where each activity is marked with its duration. Two admissible network realizations w and w' are illustrated by normal and broken arrows, respectively. The activation times t_j^w and $t_j^{w'}$ of nodes j for the realizations w and w', respectively, and the earliest possible activation times t_j^e are shown in Table 7.5.1. Note that nodes 2 and 4 are not activated in realization w and that $t_4^e=5>\bar{t}_4=3$. If we have $\chi_1^+=|S(1)|=2$, then the network realization $\bar{w}$ with $E_{\bar{w}}:=E_w \cup E_{w'} =E$ is admissible and activates node 4 at time $t_4^e=\bar{t}_4=3$.

When the nodes of a decision network N that satisfies (7.5.9a) are topologically ordered and $V=\{1,\ldots,n\}$, the earliest possible times of activation t_j^e can easily be computed by evaluating the optimality equation

(7.5.10) $t_j^e=\min\{t\geq 0|$ there are χ_j^- different $i\in P(j)$ such that $t_j^e+d_{ij}\leq t\}$

successively for $j=2,3,\ldots,n$ beginning with $t_1^e=0$ (note that node 1 is the source). The time complexity of this procedure is $0(|E|)$ because each arc need only be traced once.

If, in addition to (7.5.9a), all nodes $j\in V\backslash\{r\}$ have IOR entrance $(\chi_j^-=1)$, then (7.5.10) reduces to

$$t_j^e = \min_{i\in P(j)} (t_i^e+d_{ij})$$

That is, t_j^e equals the length of a shortest path in N from the source to node j. If, in addition to (7.5.9a), all nodes $j\in V\backslash\{r\}$ have AND entrance $(\chi_j^-=|P(j)|)$, then

$$t_j^e = \max_{i\in P(j)} (t_i^e+d_{ij})$$

That is, t_j^e equals the length of a longest path in N from the source to node j.

Finally, we want to rewrite optimization problem (7.5.1) in a more detailed form that corresponds to (7.2.3). Consider times t_i $(i\in V)$ which satisfy the constraints

(7.5.11) $\begin{cases} (t_j-t_i-d_{ij})w_{ij}\geq 0 & (<i,j>\in E) \\ t_i\geq 0 & (i\in V\backslash\{r\}) \\ t_r=0 \end{cases}$

for some $w\in\mathcal{E}^{\times}$. Obviously, the activation times t_i^W fulfill (7.5.11). The earliest possible activation times t_i^e satisfy (7.5.11) for some appropriate "minimal" $w\in\mathcal{E}^{\times}$ (which can be constructed as in the proof of theorem (7.5.7)). Observing (7.5.5) and theorem (7.5.7), the detailed form of problem (7.5.1) then reads as follows:

(7.5.12)
$$
\begin{aligned}
&\text{Minimize} \quad \max_{<i,j>\in E}\{(t_i+d_{ij})w_{ij}\} \\
&\text{subject to} \quad (7.5.11) \\
&\qquad\qquad \sum_{k\in P(i)} w_{ki} \geq \bar{\chi}_i u_i \quad (i\in V\setminus\{r\}) \\
&\qquad\qquad \sum_{j\in S(i)} w_{ij} \leq \chi_i^+ u_i \quad (i\in V\setminus S) \\
&\qquad\qquad w_{ij}\in\{0,1\} \quad (<i,j>\in E) \\
&\qquad\qquad u_i\in\{0,1\} \quad (i\in V\setminus\{s\}) \\
&\qquad\qquad u_s = 1
\end{aligned}
$$

In comparison with (7.2.3), we have the additional constraints (7.5.11) and the objective function is more complicated. Moreover, in addition to the node variables u_i ($i\in V$) and arc variables w_{ij} ($<i,j>\in E$), the times t_i ($i\in V$) are variables of optimization problem (7.5.12).

To solve problem (7.5.12) it suggests itself to use the same branch-and-bound algorithm as in section 7.2. The minimization problem corresponding to (7.2.4) is

(7.5.13)
$$
\begin{aligned}
&\text{Minimize} \quad \max_{<i,j>\in E}\{(t_i+d_{ij})w_{ij}\} \\
&\text{subject to} \quad (7.5.11) \\
&\qquad\qquad \sum_{k\in P(i)} w_{ki} = \bar{\chi}_i \quad (i\in A\setminus\{r\}) \\
&\qquad\qquad \sum_{j\in S(i)} w_{ij} \leq \begin{cases} 0 & \text{for } i\in\bar{A}\setminus S \\ \chi_i^+ & \text{for } i\in(V\setminus\bar{A})\setminus S \end{cases} \\
&\qquad\qquad w_{ij}\in\{0,1\} \quad (<i,j>\in E)
\end{aligned}
$$

If "feasible" times $t_i, i \in V$ (which satisfy (7.5.11) together with appropriate feasible $w_{ij}, <i,j> \in E$), are known, problem (7.5.13) reduces to a generalized min–max matching problem for the network $N_{A', A''}$ in Fig. 7.2.1 with weights $t_\alpha + d_{\alpha\beta}$ ($\alpha \in A'$, $\beta \in A''$), which can be solved by an appropriately modified min–max matching algorithm (compare Lawler (1976), section 5.7). As we have seen, if $\chi_i^+ = |S(i)|$ for all $i \in V \backslash S$, even "optimal" times t_i^e can be found by evaluating the optimality equation (7.5.10). In the general case, however, it is not known whether problem (7.5.13) is "easy" or NP–hard (compare Siedersleben (1982)).

References

Adolphson, D. and Hu, T.C. (1973), Optimal Linear Ordering, *SIAM J. Appl. Math.* *25*, 403–423

Aho, A.V., Hopcroft, J.E., and Ullman, J.D. (1983), *Data Structures and Algorithms*, Addison–Wesley, Reading, Mass.

Apostol, T.M. (1974), *Mathematical Analysis*, Addison–Wesley, Reading, Mass.

Arisawa, S. and Elmaghraby, S.E. (1972), Optimal Time–Cost Trade–Offs in GERT Networks, *Management Sci. 11*, 589–599

Baker, K.R. (1974), *Introduction to Sequencing and Scheduling*, John Wiley & Sons, New York

Bertsekas, D.P. (1976), *Dynamic Programming and Stochastic Control*, Academic Press, New York

Bruno, F.L. (1976), Scheduling Algorithms for Minimizing the Mean Weighted Flow–Time Criterion, in Coffman, E.G., Jr. (Ed.), *Computer and Job–Shop Scheduling Theory*, John Wiley & Sons, New York, 101–137

Bücker, M. (1988), Minimierung der maximalen erwarteten Verspätung in EO–Netzplänen, *Oper. Res. Proceedings 1987*, Springer, Berlin, 502–509

Bücker, M. (1990), Ein–Maschinen–Scheduling–Probleme mit stochastischen Reihenfolgebeziehungen unter besonderer Berücksichtigung polynomialer Algorithmen, *Ph.D. Thesis*, University of Karlsruhe

Bücker, M. and Neumann, K. (1989), Stochastic Single–Machine Scheduling to Minimize the Weighted Expected Flow–Time and Maximum Expected Lateness Subject to Precedence Constraints Given by an OR Network, *Report WIOR–361*, Inst. für Wirtschaftstheorie und Oper. Res., University of Karlsruhe

Bücker, M., Neumann, K., and Rubach, T. (1990), Algorithms for Single–Machine Scheduling with Stochastic Or Precedence Relations to Minimize Weighted Expected Flow–Time or Maximum Expected Lateness, submitted to publication

Chvátal, V. (1983), *Linear Programming*, W.H. Freeman, New York

Cinlar, E. (1975), *Introduction to Stochastic Processes*, Prentice–Hall, Englewood Cliffs

Crowston, W. and Thompson, G.L. (1967), Decision CPM – A Method for Simultaneous Planning, Scheduling and Control of Projects, *Oper. Res. 15*, 407–426

Delivorias, P.N. (1979a), Früheste und späteste Termine bei GERT–Netzplänen, *Discussion Paper No. 126*, Inst. für Wirtschaftstheorie und Oper. Res., University of Karlsruhe

Delivorias, P.N. (1979b), Früheste und späteste Termine bei GES–Netzplänen, *Discussion Paper No. 127*, Inst. für Wirtschaftstheorie und Oper. Res., University of Karlsruhe

Delivorias, P.N. (1990), Die Kostenplanung von Projekten mit Hilfe von steuerbaren GERT–Netzplänen, *Ph.D. Thesis*, University of Karlsuhe

Delivorias, P.N., Neumann, K., and Steinhardt, U. (1984), Gradient–Projection and Policy–Iteration Methods for Solving Optimization Problems in STEOR Networks, *ZOR 28*, 67–88

Denardo, E.V. (1967), Contraction Mappings in the Theory Underlying Dynamic Programming, *SIAM Review 9*, 165–177

Denardo, E.V. (1982), *Dynamic Programming*, Prentice–Hall, Englewood Cliffs

Dieudonné, J. (1969), *Foundations of Modern Analysis*, Academic Press, New York

Elmaghraby, S.E. (1977), *Activity Networks: Project Planning and Control by Network Models*, John Wiley & Sons, New York

Fix, H.P. (1979), Auswertung allgemeiner GERT–Netzpläne, *Diploma Thesis*, Inst. für Wirtschaftstheorie und Oper. Res., University of Karlsruhe

Foulds, L. and Neumann, K. (1989), Temporal Analysis, Cost Minimization, and the Scheduling of Projects with Stochastic Evolution Structure, *Asia Pacif. J. of Oper. Res. 6*, 167–191

Garey, M.R. and Johnson, D.S. (1979), *Computers and Intractability*, W.H. Freeman and Comp., San Francisco

Gittins, J.C. (1989), *Multi–Armed Bandit Allocation Indices*, John Wiley & Sons, New York

Heine, H. (1979), Numerische Behandlung von GERT–Netzplänen, *Diploma Thesis*, Inst. für Wirtschaftstheorie und Oper. Res., University of Karlsruhe

Heyman, D.P. and Sobel, M.J. (1984), *Stochastic Models in Operations Research*, *Vol. II*, McGraw–Hill, New York

Hindelang, T.J. and Muth, J.F. (1979), A Dynamic Programming Algorithm for Decision CPM Networks, *Oper. Res. 27*, 225–241

Hinderer, K. (1970), *Foundations of Nonstationary Dynamic Programming with Discrete Time Parameter*, Lecture Notes in Oper. Res. and Math. Systems, Vol. 33, Springer, Berlin

Horn, W.A. (1972), Single–Machine Job Sequencing with Treelike Precedence Ordering and Linear Delay Penalties, *SIAM J. Appl. Math. 23*, 189–202

Lawler, E.L. (1973), Optimal Sequencing of a Single Machine Subject to Precedence Constraints, *Management Sci. 19*, 544–546

Lawler, E.L. (1976), *Combinatorial Optimization: Networks and Matroids*, Holt, Rinehart and Winston, New York

Lawler, E.L. (1979), Efficient Implementation of Dynamic Programming Algorithms for Sequencing Problems, *Report BW 106*, Mathematisch Centrum, Amsterdam

Lawler, E.L., Lenstra, J.K., and Rinnooy Kan, A.H.G. (1982), Recent Developments in Deterministic Sequencing and Scheduling: A Survey, in Dempster, M.A.H., Lenstra, J.K., and Rinnooy Kan, A.H.G. (Eds.), *Deterministic and Stochastic Scheduling*, D. Reidel, Dordrecht, 35–73

Lenstra, J.K. and Rinnooy Kan, A.H.G. (1978), Complexity of Scheduling under Precedence Constraints, *Oper. Res. 26*, 22–35

Lenstra, J.K., Rinnooy Kan, A.H.G., and Brucker, P. (1977), Complexity of Machine Scheduling Problems, *Ann. Discrete Math. 1*, 343–362

Moder, J.J., Phillips, C.R., and Davis, E.W. (1983), *Project Management with CPM, PERT, and Precedence Diagramming*, Van Nostrand Reinhold, New York

Nemhauser, G.L. and Wolsey, L.A. (1988), *Integer and Combinatorial Optimization*, John Wiley & Sons, New York

Nemhauser, G.L., Rinnooy Kan, A.H.G., and Todd, M.J., Eds. (1989), *Optimization*, Handbooks in Oper. Res. and Management Sci., Vol. 1, North–Holland, Amsterdam

Neumann, K. (1977), *Operations Research Verfahren, Band II*, Carl Hanser, München

Neumann, K. (1979), Recent Advances in Temporal Analysis of GERT Networks, *ZOR 23*, 153–177

Neumann, K. (1983), GERT Networks with Several Sources, *Report WIOR–164*, Inst. für Wirtschaftstheorie und Oper. Res., University of Karlsruhe

Neumann, K. (1984a), An Optimality Equation for Stochastic Decision Networks, *Wiss. Zeitschrift Techn. Hochschule Leipzig 8*, 79–87

Neumann, K. (1984b), Precedence Relation and Stochastic Evolution Structure of General Projects, *Methods of Oper. Res. 48*, 399–422

Neumann, K. (1984c), Reducible GERT Networks, *Report WIOR–165*, Inst. für Wirtschaftstheorie und Oper. Res., University of Karlsruhe

Neumann, K. (1984d), Recent Developments in Stochastic Activity Networks, *INFOR 22*, 219–248

Neumann, K. (1985a), EOR Project Networks, *Computing 34*, 1–15

Neumann, K. (1985b), Stochastic Single–Machine Scheduling with GERT–Like Precedence Constraints, *Research Report No. 85–3*, Industrial & Systems Engineering Dept., University of Florida

Neumann, K. (1987a), Graphen und Netzwerke, in Gal, T. (Ed.), *Grundlagen des Operations Research, Teil 2*, Springer, Berlin, 1–164

Neumann, K. (1987b), Netzplantechnik, in Gal, T. (Ed.), *Grundlagen des Operations Research, Teil 2*, Springer, Berlin, 165–260

Neumann, K. (1989), Scheduling of Stochastic Projects by Means of GERT Networks, in Slowinski, R. and Weglarz, J. (Eds.), *Advances in Project Scheduling*, Elsevier, Amsterdam, 467–496

Neumann, K. and Steinhardt, U. (1979a), *GERT Networks and the Time–Oriented Evaluation of Projects*, Lecture Notes in Economics and Math. Systems, Vol. 172, Springer, Berlin

Neumann, K. and Steinhardt, U. (1979b), Time Planning by Means of GERT Networks with Basic Elements, *Methods of Oper. Res. 30*, 111–129

Nicolai, W. (1980), On the Temporal Analysis of Special GERT Networks Using a Modified Markov Renewal Process, *ZOR 24*, 263–272

Nicolai, W. (1981), *Time–Oriented Analysis of Projects Using Stochastic Network Techniques*, Math. Systems in Economics, Vol. 61, Anton Hain, Königstein/Ts.

Nicolai, W. (1982), Optimization of STEOR Networks via Markov Renewal Programming, *ZOR 26*, 7–19

Papadimitriou, C.H. and Steiglitz, K. (1982), *Combinatorial Optimization*, Prentice–Hall, Englewood Cliffs

Pinedo, M. (1982), On the Computational Complexity of Stochastic Scheduling Problems, in Dempster, M.A.H., Lenstra, J.K., and Rinnooy Kan, A.H.G. (Eds.), *Deterministic and Stochastic Scheduling*, D. Reidel, Dordrecht, 355–365

Pritsker, A.A.B. (1977), *Modeling and Analysis Using Q–GERT Networks*, John Wiley & Sons, New York

Pritsker, A.A.B. and Sigal, C.E. (1983), *Management Decision Making – A Network Simulation Approach*, Prentice–Hall, Englewood Cliffs

Rohatgi, V.K. (1976), *An Introduction to Probability Theory and Mathematical Statistics*, John Wiley & Sons, New York

Rubach, T. (1982), Single–Machine Scheduling in EOR Networks, *Discussion Paper No. 179*, Inst. für Wirtschaftstheorie und Oper. Res., University of Karlsruhe

Rubach, T. (1984), Stochastische Reihenfolgeplanung mit Hilfe von GERT–Netzplänen, *Ph.D. Thesis*, University of Karlsruhe

Schrage, L. and Baker, K.R. (1978), Dynamic Programming of Sequencing Problems with Precedence Constraints, *Oper. Res. 26*, 444–449

Schwarz, J. (1981), Numerische Auswertung von STEO–Netzplänen, *Diploma Thesis*, Inst. für Wirtschaftstheorie und Oper. Res., University of Karlsruhe

Siedersleben, J. (1981), Structural Questions with GERT Networks, *ZOR 25*, 79–89

Siedersleben, J. (1982), Optimierung von deterministischen und stochastischen Entscheidungsnetzplänen, *Ph.D. Thesis*, University of Karlsruhe

Smith, W.E. (1956), Various Optimizers for Single–Stage Production, *Naval Res. Logist. Quart. 3*, 59–66

Stoer, J. and Bulirsch, R. (1980), *Introduction to Numerical Analysis*, Springer, New York

Whitehouse, G.E. (1973), *Systems Analysis and Design Using Network Techniques*, Prentice–Hall, Englewood Cliffs

Whittle, P. (1982), *Optimization Over Time – Dynamic Programming and Stochastic Control, Vol. I*, John Wiley & Sons, New York

Whittle, P. (1983), *Optimization Over Time – Dynamic Programming and Stochastic Control, Vol. II*, John Wiley & Sons, New York

Wietek, D. (1983), Effizienzuntersuchungen für Algorithmen zur Bestimmung frühester und spätester Termine in GERT–Netzplänen, *Diploma Thesis*, Inst. für Wirtschaftstheorie und Oper. Res., University of Karlsruhe

Index